THE SCIENCE
OF ENERGY

THE SCIENCE OF ENERGY

PAYMAN SATTARI

THE LANGUAGE OF TRUTH - BOOK 1

PRAGDA
P R E S S

THE SCIENCE OF ENERGY. Copyright © 2023 Payman Sattari

First edition published 2023

ISBN: 979-8-9896275-0-9 (hcv)
ISBN: 979-8-9896275-1-6 (pbk)
ISBN: 979-8-9896275-2-3 (ebk)
ISBN: 979-8-9896275-3-0 (kdl)

For my younger self
Who never found the book he was looking for,
But never gave up his odyssey.

And for other youth
Whose souls yearn,
Thirsty for truth
In a world that has forgotten itself.

"The world was to me a secret which I desired to divine. Curiosity, earnest research to learn the hidden laws of nature, gladness akin to rapture, as they were unfolded to me, are among the earliest sensations I can remember.

...It was the secrets of heaven and earth that I desired to learn; and whether it was the outward substance of things or the inner spirit of nature and the mysterious soul of man that occupied me, still my inquiries were directed to the metaphysical, or in its highest sense, the physical secrets of the world."

— Mary Shelley, *Frankenstein (1818)*

CONTENTS

CHAPTER 1   The Nature of Reality   1

PART I: THE HISTORY OF TRUTH

CHAPTER 2   The Origins of Philosophy and Science   11

CHAPTER 3   Reason and Faith   19

CHAPTER 4   The Scientific Revolution   27

CHAPTER 5   The Enlightenment   37

PART II: THE SENSES AND THE INTELLECT

CHAPTER 6   The Subject and The Object   49

CHAPTER 7   The Observer   59

CHAPTER 8   Mind and Matter   85

CHAPTER 9   Feeling and The Sense of Touch   97

PART III: THE ORGANISM AND THE MACHINE

CHAPTER 10   The Ideal and The Actual   109

CHAPTER 11   The Atom and The Cell   119

CHAPTER 12   DNA and Identity   131

PART IV: THE EAST AND THE WEST

CHAPTER 13   The Whole and The Parts   143

CHAPTER 14   The Continuous and The Discrete   155

CHAPTER 15   Complementarity   165

CHAPTER 16   The Measurable and The Immeasurable   173

PART V: TRUTH

CHAPTER 17   Consciousness   183

CHAPTER 18   Bias   193

CHAPTER 19   Belief   201

CHAPTER 20   Truth   211

AFTERWORD   **221**

NOTES   225

BIBLIOGRAPHY   231

CHAPTER 1

THE NATURE OF REALITY

What is the meaning of our existence? Who are we and why are we here? These questions and others related to our consciousness and experience of the world were involved with our study of nature for almost 2,000 years. In that period, philosophy and science were bound together into one field known as *natural philosophy*. The "subjective observer," the one at the center of reality experiencing and interacting with it, still enjoyed a central position within the order of the cosmos. We cared about and examined physical phenomena too, but we as spectators and participants were not yet cut out of it.

All of this shifted during the 15th-16th centuries when a new methodology in the study of nature emerged, placing emphasis on physical phenomena, quantitative analysis, and material objects. This materially oriented approach not only reshaped our perception of reality, but also revolutionized nearly every facet of human society. Yet, in this transformation, the observer—the living subject central to all scientific observations—was curiously cut out of the picture. This paradigm cast the universe as an immense machine, relegating us to the sidelines as mere byproducts.

Half a millennia later, this approach does not seem to have solved our most pressing problems. For all the improvements it has brought to our society and the conveniences it has afforded us, we continue to war with each other (and with nature). For all our incredible technology, we continue to be plagued by collective experiences of fear, division, and worldly chaos. Sicknesses of every kind ravage our planet: sicknesses of the mind, sicknesses of the body, and sicknesses in nature. Not only have things not improved, but our technology has created new and more sophisticated ways of harming ourselves. Our societal structures and our general scientific worldview both continue to behave with a cold indifference to human, living aspects of our existence. What has caused this?

Clearly these problems are not new. All we need do is open a history book to see the countless ways we have warred with each other since the beginnings of civilization. But why? What is causing all of this chaos? In the modern world, we're able to create incredibly sophisticated harmonies between parts—like in our rocket ships and particle accelerators and supercomputers—but when it comes to human relationships, the simplest of harmonies eludes us. It's not enough to simply say "That's different," "That's unknowable," or "There are no answers" (a sentiment you find all too often even among academic and scientific works). *If we as subjective beings exist within this world of nature, then we are a part of her order.* There is a cause and effect to all of this. There is a mechanism.

The question is, how do we identify and understand this mechanism? How do we understand the inner, subjective aspects of existence with the same logical and technical rigor with which we understand physics? And how can we use this understanding to bring order to our inner lives and relationships, leading to a more peaceful human society? Clearly, technology alone is not enough. The core of our problem does not live with our intellect. The core of our problem is with our *inner experience*. The question is, how does this inner reality really work? What are the mechanics involved?

There is a long-standing and unspoken assumption that it is impossible to discover general laws and mechanics when it comes to our experience. We have fields like psychology and psychiatry, but they are not "exact" sciences like physics or chemistry. In other words, they do not follow from the cause-and-effect relationships of natural laws. Because our subjectivity—our inner experiences—are different for every observer, it seems impossible to identify universal laws.

Yet if we as subjective observers exist within this physical matrix—if we exist in reality—*then we are a part of nature.* All things in nature follow discernible patterns—cosmological laws. *There is an order.* If this order applies to objects in motion, why should it not also apply to the subject and their inner experience? *It does.* What is required is a new way of looking at things.

ONTOLOGIES

There is a word that can help us in our process—*ontology*. What is an ontology? It is usually defined in one of two ways. On the one hand, it can refer to anything at the existential level, the level of being or existence. In this sense, an ontological question is nearly identical to an existential question, like, "What is the origin of all things?" "Is reality really physical?" or "What is the nature of goodness?" etc. These are ontological questions because they are questions about the nature of reality.

On the other hand, an ontology can refer to a *system*. This type of ontology is a way of classifying all of the individual elements (or we could say "entities") that live within a field of knowledge, and the relationships between those entities. For example, if we're looking at cars, we can create a category for diesel-engine cars and another for gas-powered cars. Further, we can create other buckets to separate them by drivetrain (4WD, 2WD, etc.), by make and model, weight, acceleration, or top speed. All of these categories and classifications we choose to organize the data and identify the various nuances and differences between them form a system. This is an ontology.

When it comes to the interpretation of reality itself, this process must also happen. Reality also has certain essential elements that must be identified and classified to paint a picture of the whole structure. Since the earliest periods of history, philosophers have been developing ontologies to account for the overall cosmic order. Even modern science puts forward an ontology (even if it is without intending to). These models speak to the basic entities that come together to form reality, and the relationships between them.

THE NATURAL ORDER

The problem with modern science's ontology is that it doesn't account for everything. While it speaks to objective, object-oriented, and quantitative phenomena, it does not account for subjective experience or qualitative data. To speak to those missing elements and create a system that accounts for the totality, we must turn to another word, also often misunderstood, that stands at the heart of our discussion. *Metaphysics.*

Metaphysics is sometimes associated with the occult or supernatural. There is an idea alive in society that if we are talking about something that cannot be seen with our physical eyes, that it must be "spooky," unprovable, or out of reach (if real at all). But it doesn't have to be so esoteric. There are phenomena, easily within reach, that can demonstrate the reality of the non-physical.

Consider *experience itself*, for example. Even if you believe experience to be caused by the brain or the body, it's not something you can just point at with your finger or measure with a ruler. *Nevertheless, that doesn't mean it doesn't exist.* Experience is a real and tangible phenomenon. There is nothing "esoteric" about being a conscious being with experiences. But because we have become so accustomed to a materially oriented way of looking at things, and this phenomenon doesn't fit so neatly into that model, we use highly unscientific words like "spooky" or "mysterious" to refer to it. Conscious experience is not spooky or mysterious. *It is part of the natural order.* The study of the natural order in its entirety—including physical objects, conscious subjects, and the source and origin for both—is metaphysics.

As long as you seek a purely objective view of the cosmos, one in which you, as an observer, are not a part of it, your outlook on the real world is not likely to make a lot of sense. Cutting yourself out of reality, your own conscious being, has an effect on the way reality will seem to you. Only along this line of thought—the ontological worldview of total objectivity—would something as basic and essential as your own existence seem strange and out of place in your total assessment of reality.

Part of the problem has to do with our definition of reality itself. Because of the ideological dominance of our current scientific models and their materially oriented focus, we tend to think of "reality" and "physical reality" as the same thing. If we were aware we were doing this, it wouldn't be as much of a problem, but usually, the distinction goes unnoticed. Because of this unconscious parallelism, when speaking of "the cosmos," we are generally thinking of *a physical place*. Rarely do we consider that our own experiencing self and the experiences of all other living beings are a part of it, and just as essential to this "reality" as any material thing.

Though subjectivity is without form, that doesn't mean it is without order. It is the discernment of this natural order that is the true work of metaphysics. When we see a woman sitting with her head in her hands at a park bench, it is not her physical appearance that concerns us. What is evident through empathy that is not visible to our physical eyes is this person's *inner experience*. She may be experiencing heartache or grief and be in utter agony, but if we take her physical appearance to be the sum total of reality, this truth will not be visible to us. While we know that this outer, physical reality is governed by laws, what has not yet occurred to us is that the inner, subjective reality might also be.

Could there be *inner* laws of nature governing the *experiential* aspects of reality pertaining to our subjectivity? If we were to study such laws, what would we call it? Of course, delving into a space that is new to us runs into the problem of a lack of vocabulary to describe what we're confronted with. We're forced either to invent new words or to repurpose existing ones. I've elected for the latter. If physics is the study of the outer, material world (the woman's body and physical appearance) and the causal laws governing it, metaphysics is the study of the inner, nonmaterial world (the woman's inner experience) and the laws governing that.

In fact, based on the definition I'm putting forward, metaphysics isn't limited to the inner reality either. That is simply the aspect it is looking at that physics isn't. Metaphysics would be dealing with *the sum total of the whole order of reality*, physical and non-physical, inner and outer, conscious and material. Metaphysics studies *the nature of reality*, not the nature of *physical* reality. Therefore a "metaphysical" question is a question dealing with being, existence, and reality as a whole.

An ontological question and a metaphysical question are almost the same thing. They are both referring to the very essence of reality, the nature of things. The only difference is that an "ontology" can also refer to a system. If we combine these two words, we get a beautifully succinct way of referring to a system that defines all the basic elements of reality and the relationships between them—*a metaphysical ontology*.

METAPHYSICAL ONTOLOGY

Whether we realize it or not, we all have one of these. They don't have to be created intentionally, or even come from any kind of reasoning. At least at first, the beliefs we adopt about the nature of reality usually come from our parents and the authority figures we grow up with. Ultimately, we accept some version of things: a story of how things came to be, how we got here, who we are—and more to the point, *what is real*. The network of connected beliefs that form this interpretation is our ontology. It is our personal system of beliefs forming our interpretation of reality. Regardless of where this system comes from—whether it be logic, personal experience, or from some authority figure—it is composed of *beliefs*, ideas and concepts about the nature of reality we have deemed to be true.

It is truth that is at the heart of the metaphysical question, because "truth" implies that something exists, that it is real. When we consider the question of truth, we tend to isolate it on one of two sides. Either we think of it from the perspective of material things (the truth about our physical reality and its mechanics, like in the sciences), or we consider it from the perspective of our subjectivity (what is true about our experiences). Nevertheless, when we consider the reality we live in, these two sides are not really separate from each other. It is only through their sum total that this whole dance of reality we find ourselves in comes to be.

Practically speaking, this is part of the issue with our current metaphysics—our collective outlook on the nature of reality. The system of logic we have for the interpretation of physical phenomena (driven largely by physics) and the system of logic we have for the interpretation of subjective phenomena (driven largely by philosophy and religion) are incompatible. Therefore, in our everyday lived experience, it is often the case that these two systems of thought come into conflict with each other. This necessitates an integrated and cohesive system of logic that can address and incorporate both—everything for which we have data and forms a part of our experience. We need a bridge to bring the logical elements of our *experience* into harmony with the logical elements of our *physics*. As long as this is missing, the psychological framework that governs our beliefs about the nature of reality will be fragmented. Life may make sense within the narrow range of certain experiences, but we will be missing an understanding of the whole. This is why it is necessary to have a system, so the elements of both sides of this equation can work together.

THE HISTORY OF TRUTH

This system of logic forms a framework at the base of our thinking. Our psychology is ultimately driven by a logic tree with certain beliefs closer to the root of the tree, and other statements and branches of logic built off of them. For example, "The world is primarily material," "I am a physical body," etc. If a statement close to the root of the tree is false, then all of the logic built off of it must also be called into question. Therefore, to have a strong foundation for our ontology, we must evaluate root statements close to the base—foundational statements about what is real or what is true. We must see upon what assumptions our beliefs about the nature of reality are based. When we have a clear picture of our foundational ideas and how we arrived at them, we can more easily identify what is missing, and fill in the gaps.

So let's take a look at the evolution of our cosmology and the history of truth as we have seen and experienced it. Let's look at the core ideas we have circulated in the human vernacular to understand the nature of things. The history of truth deals primarily in three areas: philosophy, science, and religion. These three have formed the basis of our thinking about the nature of reality for over 3,000 years. It is helpful to understand this history because it is the easiest way to understand the nature of our metaphysics today and to identify the root assumptions at the base of our thinking. Our goal is to develop a *new* ontology. A system that can account for the whole cosmic order, where subjective and objective elements can meet and work together. But before we get there, we must start at the beginning.

PART I

THE HISTORY
OF TRUTH

CHAPTER 2

THE ORIGINS OF PHILOSOPHY AND SCIENCE

At every point throughout history humanity has had a general metaphysical outlook, an idea of the real world colored by our beliefs about what is real. In very ancient times, this was primarily attributed to deities who were in charge of keeping the order of things. Thunderstorms were caused by angry gods. Strokes of luck or misfortune were at their whims. It was not until the sixth century BC that we began to get a grip on our natural environment and realize that, through the observation of natural processes, we could discern patterns which could accurately predict the outcomes and behaviors of natural events.

On the west coast of Anatolia (what is now modern-day Turkey), there existed a Grecian state named Ionia. It was here in Ionia, in the city of Miletus, that the first three philosophers of the Western tradition were born. Thales—often attributed to being the first philosopher—followed by Anaximander and Anaximenes, put together ontologies that were altogether different than the ones that came before them. These were not ontologies built on the backs of gods and goddesses, deities that could change reality for us at their whim, but were built on *reason*, a sober analysis of the natural phenomena observable to us. One of the unspoken implications of their work was that *the world was comprehensible*. Prior to this, the comprehensibility or intelligibility of the natural world by human beings was not a given.

ORDER, NATURE, AND COSMOS

The metaphysics of those who came before Ionia was mostly defined by gods, myths, and rituals. Ideas were put forward to account for the nature of universal phenomena like the sun and the moon, the sky and the earth, fire, water, and air, etc. There were concepts of cosmological order, but this order was primarily considered to be held together by *subjective entities* (deities), rather than by *objective processes*. Therefore, nature was not considered to be the domain of man, but the domain of beings of a higher order. Then, in order to effect change in the physical world, rather than seeking to understand and put to use the mechanisms of natural processes, one looked to appease the entities believed to be in control of them.

In fact, the concept of "nature" itself (as we know it today) did not really exist yet. It wasn't until this period with our initial Ionian philosophers that the term began to gain traction. Their word for nature—*physis (φύσις)*—was derived from the Proto-Indo-European word root "bheue-" meaning "to be, to exist, or to grow." This is the same word root as the modern English "to be," arriving by way of the Old English "beon." The Greek *physis* was then later translated into Latin as *natura*, eventually leading to the English term "nature."

Nature is generally considered in two contexts. On the one hand, it refers to *a process*—that which occurs on its own when there is no interference from the outside. Plants grow, waves crash, wind blows, fire burns, etc. Things grow and change without any outside interference, force, or artificial guidance, according to some predetermined set of laws. There is a sort of "innate" set of guidelines that things follow to determine which way they will grow, move, or change. There is *order*. The discernment of this natural order was the work of these original philosophers, and what later became the work of the institution we call science.

The second context in which we typically consider nature speaks to something's *innate quality or characteristics*. What it is apart from other things when it is not artificially altered or intentionally fashioned into something else. Rather than this speaking to how something becomes, this speaks to *what it is*. This correlates strongly with the two definitions of the Proto-Indo-European word "bheue," in which "to be" and "to grow" are intertwined. To study the nature of something in this sense then, is to understand what it is apart from other things, what qualities and characteristics are inherent to it and belong to it uniquely. In both senses we are speaking to what is innate, but here we are speaking to what is innate to something's quality, and in the other, what is innate to the way it grows and changes. Either way, we are speaking of "nature."

Inherent to any idea of nature is that *there is order to the universe*. This natural order is what the Greeks referred to as *kósmos (κόσμος)*. This, in turn, was derived from the earlier kónsmos, coming from Proto-Indo-European "kens-" or "kems-" meaning "to put in order." The idea of a "cosmos" as opposed to the term "universe," implied that there was some sort of *universal order*. This is tied to the concept of nature and the comprehensibility of the observable universe. This universal order behaves according to certain cause-and-effect mechanisms that would later be called *natural laws*.

If the universe is possessed of a natural order we call "cosmos," then that means that all phenomena are the result of *natural causes*. In other words, the cause-and-effect relationships that lead to the outcomes we witness in the natural world would not actually be the result of *subjective entities*, but the result of *objective processes*. If we are to understand the mechanism that is behind the results and derive the natural laws, then we will be able to use our knowledge to predict outcomes and "act on" nature in intentional ways. We will be able to use our understanding of mechanism to produce specific results. This is what we call *technology*, coming from the Greek *techne (τέχνη)* meaning "art" or "craft," which in turn comes from the Proto-Indo-European word root "teks-" which means "to weave" or "to fabricate." Technology is really just the use of the understanding of natural processes to create tools. This can be as simple as a megalith to lift rocks, or as complex as a rocket to leave the atmosphere.

ORDER AND CIVILIZATION

Despite the fact that the systematic study of nature largely started with the ancient Greeks, there were still primitive forms of technology in the ancient world and a rudimentary understanding of certain basic mechanisms of nature. The Greek Enlightenment was built on the backs of the knowledge of many who came before them and with whom their sages often studied. The Sumerians and Akkadians (Assyrians and Babylonians) of Mesopotamia, the Hittites and Lydians of Anatolia, the Elamites, Persians, Parthians, and Medes of ancient Iran, and of course the many successive dynasties of the Egyptians over the course of the prior 2,000 years had already led to the development of certain basic technologies and knowledge of aspects of nature.

The Sumerians, who are considered to be among the earliest of the developed civilizations, began to order different aspects of community life. It was this aspect of *ordering* that was central to transforming what was otherwise a more tribal lifestyle into what may be called a "civilization." Even within the Egyptian cosmology, there was an ontological distinction between order *(ma'at)* and chaos *(isfet)*, forming natural opposites within the cosmos. A civilization is often defined by its level of social order, but this aspect of "ordering" can also apply to the physical phenomena with which a society interacts. As we have already established, a discernment of aspects of the natural order—the causal mechanisms of nature—can lead to the development of "technologies." This could be something as simple as a wheel or as complex as a rocket ship. Either way, they are technologies because they result from the understanding of natural laws to exploit "mechanism" and achieve a task. The Sumerians established multiple cities, a written code of law, primitive agricultural techniques, brickmaking for building, and even basic war technologies to deal with neighboring peoples—like the Akkadians, Elamites, and Egyptians—and much of this as far back as 3500 BC. It was around the time of the Sumerians that the human race's ability to create order really began to take form.

By the time of the Ionian Enlightenment almost 2,000 years later, these "ordering mechanisms," whether applied to "subjects" to establish social order or to "objects" to create technology, had vastly expanded. There were early forms of astronomy and mathematics in Egypt and Mesopotamia, as well as artisan (metallurgy, shipbuilding, war machines, etc.) and agricultural technologies, and basic forms of medicine. The difference between the technologies that were developed prior to Ionia and the type of work that the Ionians (and later philosophers) did was that in the old world, these "ordering mechanisms" were often developed reactively in response to immediate needs (such as agricultural innovation driven by the need for food or the creation of war technology prompted by self-protection), while the Ionians aimed to establish an intellectual order based on the notion that nature or "the cosmos" was intelligible *as a whole*.

THE ESSENTIAL ELEMENTS

The first philosophers (as well as many who followed) were later called *physikoi*—from the Greek *physikos (φυσικός)*—meaning "natural philoso-pher." It is important to note when we consider the meaning of this, there is often the unspoken assumption that the term "nature" refers ex-clusively to the domain of physical phenomena. In reality, "nature" refers to those things that occur on their own and without interference, whether within the realm of experience or physical matter. Our study of "nature" encompasses the study of what is *natural*, that which occurs of its own accord. There are laws that govern the movement and growth patterns of natural bodies, and these laws are the domain of the natural scientist.

Common to most of the natural philosophers was theorizing regard-ing *a fundamental substance*, based on the idea that everything in the materi-al world could be broken down into the same material constituent. That which could then be pieced back together to form the myriad substances we see in the macroscopic world. To make order out of parts, it is essen-tial to understand the connections and correlations between them, but it is also important to understand which pieces are more *fundamental* or *primary* than other ones. This facilitates the process of creating *logical order and hierarchy*, and it also facilitates the process of boiling things down to their *essential elements*. There are certain things that must come first in a logical hierarchy and exist closer to the roots of the logic tree. These are fundamental in order to explain *other* things. This process of boiling things down to their essential elements and assessing which pieces are more pri-mary in the logical order is vital to creating a reasoned cosmology.

Also fundamental to the process of creating a reasoned cosmology is the establishment of *a line of reasoning*. A "line" implies that there is some sort of order or sequence to your reasoning. Based on the truth or veracity of one reasoned statement, we can deduce another, and so forth. Through this process, one eventually establishes *a logical framework*. A logical framework is a network of reasoned statements that connect to each other to reveal some grander truth or reality. There is a relationship between a metaphysical ontology and a logical framework. They are both essentially referring to a network of ideas that together form a cosmo-logical picture. A *truth* about the nature of things. The only difference is that while some metaphysical ontologies are based on myth and ritual, *all logical frameworks are based on reason*. This would differentiate a proper metaphysics from what might be termed religion, with the necessity of a line of reasoning being the primary difference between them.

In Ancient Greece, the original term to refer to the source, origin, or root of all things was *arche (ἀρχή)*. This corresponds with what could be called the "first principle," "first cause," or the ultimate underlying substance of all things. Within our line of reasoning or logical framework, *arche* would be considered *primary*—that from which all other things emerge—and would be the only thing which itself does not have an origin. From Aristotle onwards, "arche" was referred to as the "substratum." From this substratum, all things come to be, because it is the original principle. Since it is first, it is the parent of all things, and the one thing that all existing things have in common. In this way, it would be the only constant. The *arche* would be irreducible and the substance from which all other material things are formed.

Thales of Miletus—our first historically accepted philosopher—posited that the fundamental substance, or arche, was water. His student Anaximander, on the other hand, reasoned that since water is wet, it could not give birth to fire, another of the essential elements. He extended this reasoning to all of the universally accepted elements of the time: earth, air, fire, and water, stating that since one element could decay and transform into another, they could not be fundamental. This reasoning led him to believe the arche or first principle to be *apeiron (ἄπειρον)*, that which is unbounded and infinite. The concept of *apeiron* was connected to the previous cosmological idea of the "nothing" or "Great Void," referred to in some Greek literature as "Chaos."

While many philosophers who followed theorized on the nature of the original substance, the vital difference between the old and the new was that the process of arriving at their respective conclusions was no longer based on myth and tradition but on reason. The use of reason to deduce the order of the cosmos was then collectively referred to as "philosophy," and those individuals whose work it was to engage in the pursuit of truth and knowledge, using reason to deduce natural principles and put forward cosmologies that explain observable phenomena were referred to as "philosophers."

The term philosopher is believed to have originated with Pythagoras, a student of Anaximander. It comes from the Greek *philos (φίλος)* meaning "love," and *sophia (σοφία)* meaning "wisdom," roughly translating to "the love of wisdom." The association of wisdom with philosophy and the pursuit of a perfectly balanced cosmology, "a unified system to explain everything," began around this time in the sixth century.

THE SPREAD OF NATURAL PHILOSOPHY

The systematic study of nature and the use of reason to derive its principles continued from the sixth century and spread easily to the rest of the world. The seeds of philosophy, this effort to understand the world through reason rather than simply through mythology and folklore, already existed to some degree in the ancient world. Thales and Pythagoras, as well as many sages who followed, studied extensively in other regions like Egypt and Mesopotamia, which contained their own systems, though rudimentary, of mathematics, astronomy, and medicine. It was primarily a reasoned cosmology (as opposed to one based on myths) and an analysis of reality as a whole that was missing from the belief systems of the ancient world and was developed and spread by the ancient Ionians.

Philosophy, and what was later known as "science," were essentially the same discipline in the ancient world. Science as we know it today was referred to as *natural philosophy*, differentiated from the rest of philosophy by the fact that its focus was on explanations of the natural world, rather than putting forward conclusions about other phenomena such as politics, "right living" (ethics), or aesthetics (beauty). These fields were also a part of the endeavor to use reason to come to conclusions about the nature of things but were not always tied in with cosmologies, explanations of the nature of the universe as a whole.

The Greek *physis*, which is at the heart of both the philosophical and scientific endeavors, shares in common with English the word root "to be" or "to exist" from the Proto-Indo-European "bheau." It is this concept of *being* that connects philosophy and physics (and by extension all of science), for *being* is of the question of existence. What *is*, is true. It exists. What is not true, *isn't*. It *does not* exist. This is the fundamental question of reality and the heart of the metaphysical inquiry. *What is real?* And that is where philosophy and physics—the physical and the non-physical—meet. In the nature of being we find the aim of both scientific and philosophical pursuits, the search for truth, what *is* or *what is real*.

Both philosophy and science aim to understand the nature of things, the causal mechanisms behind observable phenomena, to point to the existent thing, what *is*. Science and philosophy make existential statements about reality, also called "metaphysical" statements, implying that reality works in a certain way. While the science we know today aims purely to explain physical phenomena, or to explain the phenomenon of consciousness *through* physics, philosophy, or more generally metaphysics, aims to understand *reality as a whole*.

CHAPTER 3

REASON AND FAITH

The difference between the fields of philosophy and science, and what we call religion, is that philosophy and science are based on *reason*, whereas religion receives its truth from the teachings of a prophet or the wisdom of ancestors (tradition). This is whether there is reason to support the truth inherent to its ideology or not. For this reason, philosophy and science are functionally different than religion and must be explainable via logic, whereas religion asks for faith, either in the truth content itself, or in those who delivered it. Religious truth is typically not *reasoned* truth. This is not a knock on religion, but a clarification of some of the qualities and characteristics that differentiate it functionally from the other two main purveyors of truth and wisdom in our society, namely philosophy and science.

Starting in the same century that modern philosophy and science were birthed in Ionia, many other cosmologies broke out all over the world. Siddhartha Gautama for example, the man who would later be known as "the Buddha," was born in India during the same century. Lao Tzu, the father of Taoism, wrote the seminal text of his cosmology, the Tao Te Ching, within the same century. Not long after, the basic tenets of Confucianism were established. Of course in Ancient Greece, there also continued to be philosophers with hugely influential ideas in the centuries that followed, including Socrates, Plato, and Aristotle.

The period beginning in ancient Ionia in the sixth century BC through what is termed "the Migration Period" around 300 AD saw the birth of a plethora of hugely influential cosmologies. One branch of the Proto-Indo-Europeans with the self-designated ethnonym *arya* or *ariana* migrated into modern-day Iran and northern India, bringing with them their respective cosmologies. This eventually turned into the Iranian holy book the Avesta which gave birth to Zoroastrianism, the state religion of Persia (and sometimes considered the first monotheistic religion), as well as the Vedas, which are the precursor to modern Hinduism. These two existed as far back as the second millennium BC, with Zoroastrianism being spread through much of the known world by the first Persian empire. This was followed in the first millennium BC by Judaism, the Israelite worship of the God Yahweh being slowly codified into an official religion after the return of the Israelites from Babylon to Jerusalem. More than one thousand years later came the birth of Jeshua ben Joseph in Bethlehem, eventually giving birth to modern Christianity.

The Migration Period, occurring roughly between 300 and 800 AD, also known from the Western perspective as "the Barbarian Invasions," was a period of relative darkness in the West when it came to philosophy and science. The period was marked by a rise in religiosity and a metaphysics based more in faith than in reason. Christianity became the state religion of the Roman Empire via Emperor Constantine and the imperial capital was moved east to Byzantium. There was a sharp decline in literacy in Europe and the works of the great Western philosophers were largely lost. After the fall of the Western Roman Empire in the fifth century, many of the original Greek manuscripts were preserved in the Eastern Roman Empire (also known as the Byzantine Empire), where Greek, rather than Latin, was the dominant language.

Toward the end of the Migration Period in the eighth century, Arab caliphates began to spread their influence and power, their empire eventually reaching mainland Europe via both Iberia (modern-day Spain) and Sicily. Arabs were increasingly exposed to Greek ideas as they took control of areas heavily influenced by Hellenic cultures, such as Egypt and the Levant. While there was an initial resistance to the adoption of many of these ideas, there began an effort by early Islamic scholars to translate their works. With the transition from the Umayyad to the Abbasid Caliphate, the efforts to both collect and translate these works took an important turn.

Between the mid-eighth and the late-tenth century, a significant volume of Greek texts, including those from Aristotle, Plato, Euclid, Pythagoras, and Plotinus, were translated into Arabic. The Abbasid Caliphate, the third caliphate to succeed the Islamic prophet Mohammad and one of the largest empires in history, saw a flourishing of Islamic scholarship. The caliphate demonstrated an interest in gathering the scientific and philosophical knowledge of its conquered peoples. This culminated in the establishment of the House of Wisdom, a great library and center of learning in its newly built capital of Baghdad. Baghdad and the House of Wisdom became a center for philosophy, science, and culture in what eventually became known as the Islamic Golden Age.

Well-known scholars such as Alpharabius (Al-Farabi / الفارابي), Avicenna (Ibn Sina / ابن سينا), Averroes (Ibn Rushd / ابن رشد), and Alkindus (Al-Kindi / الكندي) wrote extensively on many topics, including metaphysics, cosmology, logic, mathematics, medicine, physics, chemistry, and ethics. The vast resources of the empire enabled significant investment in translation. Not only Greek, but texts from various other cultures including those of Persia and India were brought to the House of Wisdom to be translated into Arabic. Often referred to as "the Translation Movement," this period was hugely influential in early Islamic philosophy. Particularly the works of Aristotle (of which many commentaries were made by various Islamic scholars) were influential in the metaphysics of this period, eventually finding their way back into Western Europe via their Arabic translations.

THE INTEGRATION OF REASON AND FAITH

Until this point in history, philosophy and science, both based in reason, and religion, based on prophetic insight, faith, and tradition, were largely separate from each other. One of the significant efforts of the early Islamic scholars was the blending of the reasoned discourse of philosophy with the metaphysical revelations of religious scripture. The works of Aristotle were particularly influential in the attempts to reconcile rationalist philosophy with Islamic theology. This blending of philosophy and theology was characteristic of this period. It produced a number of commentaries in Arabic that were later influential in the establishment of the medieval school of philosophy known as Scholasticism, which attempted to combine Christian theology with the revelations of early Greek thought.

Roughly between the years 1000 and 1250 AD in the late Middle Ages, Greco-Roman thought of antiquity began to re-enter Europe. The decline of the Byzantine Empire led many scholars to flee westward back into Western Europe, at the same time that there was increased travel to the East. This led to the rediscovery of many of the original Greek texts, either held by Byzantines or in the libraries of the Arab caliphates. A significant effort to translate these works into Latin occurred during this time, especially in those areas of overlap between European and Arab cultures, such as in Iberia and Sicily. As early as the tenth century, texts were gathered, translated, and transmitted to Europe via these translation centers.

In the eleventh and twelfth centuries, many of the previously Arab occupied areas of Europe were reconquered during the period of military campaigns known as the "Reconquista." This sped up the translation process considerably, with Toledo and Sicily being centers for the transmission of old Greek ideas back into mainstream European thought. It also opened the way for Christian theologians to encounter Islamic philosophy, often by way of their commentaries on these original Greek manuscripts.

This same period also saw the establishment of the modern university. Prior to the eleventh century, education typically occurred in monasteries and cathedrals in the West, and in Islamic *madrasas* (also theological schools) in the East. Among the first universities in Europe were the University of Bologna in 1088, the University of Paris in 1150, and the University of Oxford in 1160. University education during this period consisted of *the trivium* (grammar, rhetoric, and logic) and *the quadrivium* (arithmetic, geometry, astronomy, and music). These subjects were meant to prepare students for more advanced studies in other fields like theology, law, or medicine. The reintroduction of Greek works into mainstream European thought—especially those of Aristotle—was hugely influential in catalyzing the development of universities and fostering a broader interest in education. In the heavily religious environment of the Middle Ages, this resurgence of early Greek philosophical and scientific insights played a pivotal role in the establishment of both university education and the Scholastic movement.

Scholasticism was an effort by early Christian scholars to reconcile philosophical thought and the rigors of Aristotelian logic, with religious ideology. The works of Albertus Magnus and Thomas Aquinas were particularly influential during this time, with Aquinas' work *Summa Theologica (Summa Theologiae)* being the seminary text for Catholic theological studies for many generations to come. Aquinas artfully weaved the reasoned discourse of philosophy—taking from the newly reintroduced works of Aristotle and the commentaries of the Islamic scholar Averroes (whom he referred to as "The Commentator")—with the theological doctrines and biblical revelations of the Christian faith. *The intent was to prove the viability of biblical passages through reason.*

The result was a mirror of what many of the Islamic scholars such as Avicenna had done in the East, which was to accept the value and importance of logical discourse and reason in the deduction of natural principles, and to incorporate them with what was accepted to be an even higher truth—that which was brought forward by revelation or prophetic insight. The result was an artful synthesis of religion (theology) and philosophy. Both movements, in the East and in the West, received mixed reviews. Even Aquinas (who was generally well-respected in the religious community) was condemned by the Church for a short time for attempting to explain divine revelation through logic, what was seen in some circles as heresy.

The translation of original Greek works into Latin and the recovery of Greek (and Greco-Roman) philosophy reached its peak in the early thirteenth century. Adelard of Bath, an English natural philosopher who traveled throughout Europe and the Arab world, was pivotal in the translation of certain key astronomical and mathematical texts, including Euclid's renowned work *Elements*. Euclid was a Greek mathematician from Alexandria in Ptolemaic Egypt. His work *Elements* (produced in 300 BC) was largely lost to Western Europe for nearly a thousand years until Adelard recovered and translated it from Arabic to Latin around 1120. The first printed edition appeared a few hundred years later in 1482, after the invention of the printing press. Since then it has had more printed editions than nearly any other book in history.

Euclid's *Elements* consists of 13 books, primarily focused on plane and solid geometry, number theory, and mathematics. The logical precision of *Elements* and the way the information is presented—via organized axiomatic laws that build on each other to reveal grander truths—was part of what made this work so influential. It was used as a standard textbook in universities all the way until the twentieth century, its logical rigor and organization being unsurpassed for generations, and deeply influenced many of the thinkers responsible for the intellectual revolution of the Early Modern Period. Following the end of the Middle Ages, the Early Modern Period—roughly between 1450 and 1815—saw a radical transformation of our metaphysical outlook. This era contained three of the most influential periods in history: the Renaissance, the Scientific Revolution, and the Enlightenment.

THE BUILD-UP TO SCIENCE AS WE KNOW IT

For much of our history leading up to the early enlightenment of the Ionian philosophers, our view of the nature of reality, the metaphysical system we used to interpret our experience, was based on mythology and folklore. If the period from 600 BC to 300 AD marked a shift from an outlook based on myth to one based on reason, the period between 300 through roughly 1000 shifted the focus to faith and religion. When Greek thought was introduced into Eastern Islamic discourse and then reintroduced back to Western European thought, there was an attempt on both sides to integrate reason and faith. The reasoned discourse and logical interpretations of nature dominant in Greek antiquity were synthesized with the metaphysical assertions of religion - namely Christianity, Islam, and Judaism.

This synthesis of philosophy and religion—both fields being centered around the nature of truth and reality—heavily influenced the establishment of educational institutions, including Islamic *madrasas* and European universities. These centers of learning then became ideally suited to receive and incorporate new advances in knowledge, much of which was now more widely available due to the invention of the printing press in the mid-1400s.

Heavy emphasis on the use of reason and logic to advance human knowledge was primed by this period, between 1000 AD and the beginning of the Renaissance in the early 1400s. By the end of this period, the translation of early Greek writings from Arabic to Latin was mostly finished, and the ideas contained therein were built upon and developed by Renaissance thinkers. The gradual shift away from religious thought with the reintroduction of Greek works shifted the emphasis of European thinking away from faith and towards reason. This shift coincided with an increase in centralized institutions of learning around the world and an explosion of interest in those fields emerging from a reason-based interpretation of nature, which began in earnest with the ancient Greeks and continued with Islamic and Christian scholars. All of these factors, including the gradual disintegration of religious thought from the reasoned discourse of philosophy and an emphasis on new methods of deduction, culminated in the period known as the Scientific Revolution of the sixteenth and seventeenth centuries.

CHAPTER 4

THE SCIENTIFIC REVOLUTION

During the sixteenth and seventeenth centuries, two schools of philosophy were born that helped shape a revolution in our general cosmological outlook for generations to come: *empiricism* and *rationalism*. While these two schools of thought are not necessarily at odds with each other (one can be a rationalist and an empiricist at the same time), they differ in the degree of emphasis they place on one of two sides of a polarity. On the one hand, we have the senses and sensory experience, the information that comes to us *empirically* through experience. On the other hand, we have the mind or intellect, where associations and correlations are made between symbols and sensory experiences. The intellect is the center of logic and reason. It is what makes *order*.

Both empiricists and rationalists of this period agreed that these two come together to form our reality. What they did not agree on was which one came first in order, which was more primary in the discernment of truth. There were three prominent philosophers on one side of the argument and three on the other: Bacon, Locke, and Hume on the side of the empiricists, those leaning more heavily on the senses and experience, and Descartes, Spinoza, and Leibniz on the side of the rationalists, those leaning more heavily on the mind and intellect. All of them articulated their philosophies in the roughly two centuries that constituted the Scientific Revolution and the Enlightenment, and highly influenced both our cosmological outlook, and the methodologies that would later form the basis of the scientific enterprise.

Roughly speaking, the empiricists believed in the primacy of the senses in the determination of truth. They rightly pointed out that everything begins and ends with experience, that all of the information we have about the world is gathered through the senses. Generally, they believed that our concepts and knowledge come to us *a posteriori*, meaning directly from (or "after") experience. Rationalists on the other hand, believed in the primacy of the mind and reason in the derivation of truth, rightly pointing out that the information that reaches us through the senses can often be skewed. Further, that information can sometimes be obtained independently of the senses, like that brought through deductive reasoning, in fields like mathematics or through abstract thought. Rationalists tended to believe that our concepts and knowledge come to us *a priori*, meaning independently of (or "before") experience. The truth is, both camps—empiricist and rationalist—contributed greatly to the development of science as we know it today.

This period is often known as "The Age of Reason," not because it was against the senses or sensory knowledge, but because it was a stark deviation from the religious thought and social hierarchies based on heredity and social position characteristic of the periods that preceded it. During this period, both empirical (sense-based) and rational (intellect-based) principles came together to form a new method of inquiry into nature. *This method put trust in human cognitive faculties rather than basing truth on any source external to ourselves.* This was a serious change in the way we treated our search for truth, because it placed the impetus on us, rather than any outside source, to confirm or refute any truth statement. Inevitably, this led to a confrontation with the authoritative bodies of the time.

BRILLIANT MINDS, REVOLUTIONARY THOUGHTS

Francis Bacon was an English philosopher and scientist from the late 1500s and one of the original empiricists. He strongly advocated for experiment and empirical evidence in the study of nature as well as a more organized method of investigation. By way of experiment, one could test hypotheses by simulating specific conditions and observing and recording the results. This method emphasized the centrality of *empirical observation* in verifying any hypothesis. To make repeated empirical observations and use that data to build up to more generalized conclusions is called *inductive reasoning,* of which Bacon was a huge advocate. Bacon is sometimes referred to as the father of the scientific method, though there were others who also contributed greatly along the way.

One notable example predating Bacon by about a hundred years—and often credited with kicking off the Scientific Revolution in the first place—is Nicolaus Copernicus, a Prussian astronomer and mathematician who put forward a heliocentric model of the universe. While this may seem trivial from the modern perspective, the idea that the sun and not the Earth was at the center of the universe—a *heliocentric* rather than *geocentric* model—went against thousands of years of conventional wisdom and religious ideology. His seminal work *On the Revolutions of the Heavenly Spheres (De revolutionibus orbium coelestium)* was a compendium of six books outlining his model and was published during the year of his death in 1543. Copernicus was hesitant to publish, given that such a controversial idea might have put him at odds with the Church. After his death, his book was banned by the Catholic Church for nearly 300 years until it finally removed its ban on books related to heliocentrism in 1835.

Using many of the geometric principles outlined in Euclid's recently published work *Elements*, Copernicus was able to deduce the most accurate astronomical model of our solar system to date. He correctly identified the central position of the sun, the order of the known planets, a nearly accurate description of their orbital periods, the rotation of the Earth on its axis, and the effect of that rotation on the changing of the seasons. The reception to Copernicus' ideas was lukewarm, with some sharply condemning his views, so radically different than those which were historically accepted. But there were other astronomers and mathematicians around Europe who drew inspiration from and expanded on his work.

One such scientist was Johannes Kepler, a German mathematician and astronomer famous for his laws of planetary motion. Using Copernicus' work as a foundation, Kepler mathematically corrected many of its finer points. For example, he demonstrated that the planetary orbits around the sun were not circles but *ellipses*, with the sun at one of two focal points. Further, he showed that as a planet moves along this elliptical path, a line segment connecting it to the sun sweeps out equal areas of space during equal intervals of time. This showed that the planet's speed in its orbit changes, being faster when closer to the sun and slower when farther away. Kepler discovered these laws in the early 1600s, improving on Copernicus' model and setting the stage for other intellectual developments to come.

An interesting characteristic of this time was that many of the leading figures who would contribute most significantly to the emerging field of science were also active metaphysicians and philosophers, proposing entire systems of interpretation for the nature of reality. They often put forward comprehensive frameworks that touched upon multiple areas of human understanding, including epistemology (the study of knowledge itself and how to determine what is true), metaphysics (the study of reality as a whole), ethics (right action), politics (right governance), and physical science.

One such figure was René Descartes, a seventeenth-century rationalist philosopher and scientist from France. Despite being most commonly known for his famous phrase *cogito ergo sum ("I think, therefore I am")*, what is often less discussed is that Descartes was also a gifted mathematician and contributed significantly to the field. His development of analytical geometry—a way of studying geometry algebraically—paved the way for the later development of calculus. The Cartesian coordinate system (the grid of horizontal and vertical lines beginning at a point of origin [0,0] used in many scientific fields) was also developed by and named after him (the word "Cartesian" coming from "Cartesius" or "Cartesianus," the Latinized form of the name Descartes).

Philosophically speaking, Descartes is most well known for his development of "Cartesian dualism," or the dualism of body and mind. This theory states that while there is a field of physical matter where all material things exist and which can be measured and analyzed, there is also another field of reality called "mind" which is outside of physical space but which is connected to and interacts with it. The dualism of mind and body, and the idea that "mind" should exist as a fundamentally different space was a new and exciting concept that generated much discussion in years to come. Descartes' ingenious and novel ways of thinking were characteristic of the many intellectual developments of this century and the brilliant minds that defined it.

THE SCIENTIFIC METHOD AND QUANTITATIVE FOCUS

Another scientist influenced by the Copernican system and hugely influential in the advancement of intellectual thought after the Renaissance was Italian physicist and astronomer Galileo Galilei. Active at virtually the same time as Francis Bacon and Johannes Kepler, Galileo was one of the most significant catalysts to the development of modern science as we know it. He was a huge proponent of empirical methods of study and mathematically derived principles. Previously, many efforts to understand the natural world were focused on *qualitative* rather than quantitative principles—influenced greatly by Aristotelian physics—and were not always tied in with experiment or empirical evidence. Galileo's derivation of natural principles through mathematics and his emphasis on quantitative rather than qualitative methodologies contributed greatly to the establishment of the scientific method as the primary method of inquiry into the study of nature.

Galileo made significant contributions to many fields of knowledge, including astronomy, mathematics, and physics. The original telescope, invented by Dutch eyeglass maker Hans Lippershey (though several others applied for a patent around the same time), was able to amplify light at a rate of around three times. Upon hearing about this Dutch invention, Galileo was able to recreate his own (even without seeing one in person). He improved on the original model by learning how to grind lenses, eventually resulting in a telescope ten times more powerful than the original at a magnification rate of thirty times. With this technology, he made astronomical observations that were previously impossible deducing many new data about the nature of space and the cosmos. One such discovery was the number of stars in the sky. Previously, it was believed that what could be seen with the naked eye was the limit, but the invention of the telescope made it clear that there were vastly more. Additionally, Galileo discovered four of the moons revolving around the planet Jupiter, the phases of the planet Venus, that the moon's surface is rough rather than smooth, and dark spots on the surface of the sun.

Galileo's emphasis on quantitative methods of analysis led to the discovery of other natural principles. For example, contrary to the common belief that heavier objects fall faster than lighter ones, Galileo demonstrated that, in the absence of air resistance, all objects fall at the same acceleration due to gravity. This physical principle, among many others, was the result of a shift from qualitative to quantitative forms of study, which firmly established mathematics as the definitive methodology for the discovery of cosmological principles. Aristotelian physics—which was the dominant paradigm in Europe for many years, and relied more on qualitative rather than quantitative interpretations of reality—was slowly being eroded and replaced by the quantitative and empirical methods championed by the many intellectual scholars of this period.

Despite the advances in logic and reason put forward by Galileo, his evidence of the truth of heliocentricism was fiercely opposed by the Catholic Church. He was tried by the Inquisition in 1615 and put on house arrest, where he remained until his death in 1642 (he was not vindicated by the Vatican until 350 years later in 1992). The year of his death was the same year that another of the most influential figures in the history of philosophy and science was born, Sir Isaac Newton.

NEWTONIAN MECHANICS AND
THE BIRTH OF MODERN SCIENCE

Newton's seminal work *Mathematical Principles of Natural Philosophy (Philosophiæ Naturalis Principia Mathematica)* is universally recognized as one of the most influential works in the history of science. It was a culmination of the many intellectual advancements made by the numerous philosophers, scientists, and mathematicians who followed the reintroduction of ancient Greek thought back into mainstream European thinking. Often referred to as simply *the Principia* and composed of three books, this work synthesized and expanded upon the many mathematical models set out by the scholars who preceded him, including Kepler, Galileo, Copernicus, Euclid, and others.

While Isaac Newton contributed to many fields of science, *the Principia*, first published in 1687, established *classical mechanics*, the basic physical laws governing the motion of macroscopic bodies. This includes three laws of motion which describe the relationship between the motion of physical objects and the forces acting on them. The first law indicates that an object will remain at rest or in motion at a constant velocity unless it is acted upon by an outside force. This was a refinement of observations already made by Galileo, that objects tend to resist changes in motion, and what Kepler called *inertia*. Appropriately, Newton's first law is often called *the law of inertia*. The second law states that the *force* of an object is equal to its mass times its acceleration. And finally, the third law states that when an object exerts force on another object, the second object exerts a force equal in magnitude and opposite in direction back on the first. These three laws together laid the foundations for what is commonly referred to as *Newtonian mechanics*.

Since Newtonian mechanics describe the laws governing the behavior of physical objects, they pertain to the area of natural science we call *physics*. In physics, Newtonian mechanics are the earliest form of what is called *classical mechanics*. These are the mechanics governing the motion of virtually everything we can see. When it comes to objects smaller than an atom or objects whose velocity is so fast that they approach the speed of light, different forms of mechanics are used (quantum mechanics and special relativity respectively, though these were not established until much later).

In addition to the establishment of classical mechanics, Newton also established *the law of universal gravitation*. This was one of the first times laws governing the motion of objects on Earth (like an apple falling towards the ground for example) were connected to laws governing the behavior of cosmic bodies (like the motion of the sun or the moon). Firstly, he was able to mathematically demonstrate that the force of gravity is dependent on distance. Specifically, that the force of gravitational attraction between two bodies is inversely proportional to the distance between them (technically, at their centers). Secondly, given that the third law of motion indicates that an object exerting force on another object receives a force equal in magnitude back in the opposite direction, an object affected by another object's gravity must also be affected by its mass in addition to its distance. These two principles came together to form Newton's law of universal gravitation, that the force of gravitational attraction between two bodies is dependent on both of their masses and inversely proportional to the square of the distance between their two centers (if that went over your head, don't worry, it's not critical for our discussion). Basically, this means that more massive bodies will exert a greater gravitational force on each other, and that objects more distant from each other will exert a weaker attraction. What made this law extra special was that it was applicable *universally*. That means that *all bodies exert gravitational attraction*, whether cosmic (off-Earth) or terrestrial (on-Earth).

Newton's work in physics and geometry also led to significant developments in mathematics. At virtually the same time as Gottfried Leibniz—a German logician, mathematician, and rationalist philosopher—Newton developed *calculus*. If geometry is the study of shapes, and algebra is the study of generalized arithmetic operations, calculus is *the mathematical study of change and motion*. Therefore, it is fundamentally related to physics. Originally called "infinitesimal calculus," this form of mathematics made it possible to do math with objects in motion, whereas the mathematics of the Greeks was primarily made for static bodies. This meant that physical systems could be analyzed with new quantitative methods, making it possible to draw more precise calculations of motion and change, and hence derive new laws. Between his three laws of motion, the principle of universal gravitation, and the new mathematics of calculus, all the tools necessary to study the mechanical relations of ordinary physical bodies were firmly established.

In addition to laying the foundations for modern physics and mathematics, Newton also made significant advances in the field of optics (the branch of physics studying the behavior and properties of light). The telescope, developed by Lippershey and improved upon by Galileo, was further refined by Newton. Up until this point, there had already been research into the use of mirrors rather than lenses in the construction of telescopic instruments, but Newton was the first to successfully build one. Often called a reflecting telescope (and subsequently a Newtonian telescope), he built the first functional model in 1668. The results of his findings on his research into light were summarized in his other highly influential work *Opticks*, published in 1704. *Opticks* laid the foundations for further work into the fields of heat, light, magnetism, and electricity, and further strengthened the movement toward empiricism, as it heavily emphasized the necessity for experimental evidence in the establishment of scientific conclusions. *Opticks* made several advances in thought, including that "pure" light (like sunlight, for example) was not purely white or colorless, but was in fact composed of a spectrum of color revealed by refracting it through a prism.

THE INSTITUTIONALIZATION OF KNOWLEDGE

The explosion of new research during the Scientific Revolution, as well the growing university system, created the necessity for better organization of the flurry of new ideas and concepts generated by the increasing community of scientific scholars. In the early seventeenth century, *scientific societies* began to pop up all over Europe, culminating in two of the most influential of these institutions of the time: the Royal Society of London for the Promotion of Natural Knowledge created in 1662, and the Académie des Sciences of Paris in 1660. In these societies, natural philosophers from all over Europe could meet, share research, debate new ideas, discuss their findings, and collaborate. Our collective knowledge and inquiry into truth and reality were becoming more and more centralized and institutionalized as this period moved on.

The sheer volume of new information being generated at this time, especially after the publishing of Newton's *Principia*, created the need for greater organization of data and easier ways of making new information available to the community. This led to the first scientific journals for the publishing of scientific papers and the dissemination of new knowledge. Between the scientific societies rapidly spreading throughout Europe, the newly minted university system, and these new scholarly journals, an infrastructure was being set in place that would lead to an explosion of new knowledge from a much wider community of academics and scholars. Coinciding with increased literacy in Europe and greater public interest in the sphere of natural knowledge, the Scientific Revolution of the seventeenth century set the stage for the highly transformative period to follow known simply as "The Enlightenment."

CHAPTER 5

THE ENLIGHTENMENT

The Enlightenment and the period immediately preceding it are often collectively referred to as "The Age of Reason." It is the use of reason in the discernment of natural principles, but also best principles in virtually any other field, that is the mark of all philosophy. For the roughly 700 years between 300 and 1000 AD in the period sometimes called the "Dark Ages," Western philosophy was all but lost to Europe. With the return of Greek philosophy to the West during the Translation Period, the reasoned discourse of philosophy, and faith in humanity's innate faculties like sense and reason, were slowly restored. This culminated in the Scientific Revolution, where a number of brilliant minds built upon the recovered knowledge of the Greeks and established a new and more effective method of inquiry into nature.

While this new method caused a massive shift in our ontological outlook, it also greatly affected our view of our own role within it. One of the basic ideas implicit in ancient Ionia—and in all philosophy and science—is that *the world is comprehensible*. This means that by using *human faculties alone*, one can comprehend nature and use that knowledge to achieve material outcomes. This was at odds with the idea that nature was the domain of entities of a higher order than man. That one must kneel at the altar of a deity, or a priest or king who represents that deity, so they can achieve those outcomes for you. While in the early days of philosophy, this kind of reliance on one's own faculties was less of a problem, in the deeply religious and aristocratic environment of Enlightenment-era Europe, that was not the case.

ENLIGHTENMENT-ERA EUROPE - SETTING THE STAGE

The landscape of Europe at the time leading up to the Enlightenment was marked by the political and ideological dominance of the Church, monarchy, and aristocracy. Social classes were highly stratified, with the aristocratic and religious elite holding a position of advantage in virtually every area of life, from land holdings to governance to treatment under the law. An explosion of newly published materials (made possible in part by the printing press), increased literacy in Europe, and a burgeoning network of scientific societies and universities resulted in a better-informed public with more access to information. This increased access to information, coupled with new social philosophies based on reason, encouraged greater reliance on one's own faculties rather than blind deference to an external authority. This began to cause cracks in this structure, encouraging greater independence of thought among common people.

In 1689, English physician and philosopher John Locke put forward a highly popular treatise in support of empiricism called *An Essay Concerning Human Understanding*. In this work, Locke argued that every human mind at birth is a blank slate (what he called "*tabula rasa*") that is imprinted on by sensory experiences as we grow up (and which we can then reflect on to gain new knowledge). He argued against the idea that any knowledge is "innate," and that the only real knowledge we can gain about the world comes either through experience or reflection. In order to have a more enlightened outlook then, one must treat themselves as what he called an "*ignoramus*," one who is without any knowledge, obligation, or servitude at birth. He further argued that accepting this basic ignorance returns to us our liberty and agency to act freely. The idea that we are born free of any of these pre-programmed realities was also present in his political philosophy, which he outlined in his 1690 work *Two Treatises of Government*.

The ideas presented in *Two Treatises* are often credited with beginning the wave of revolutionary sentiment that led to the massive social and political changes of this period. Central to this philosophy is something called "*state of nature*," a popular philosophical concept at the time that imagines what man would be like before organized civil societies and governments were formed. Locke argues that in our "state of nature" we are free to organize our thoughts, behaviors, and choices as we please, and that in this state, the only laws that limit us are the laws of nature themselves.

Implicit in these concepts is that all people are born free and equal. This idea opposes the notion that a god or deity made them naturally subject to the rule of a monarch—what is often referred to as "the divine right of kings." Invoking the laws of nature as grounds for justifying the equality of all people (including those born into nobility or monarchy) was a groundbreaking and emotionally powerful argument for the time. This was especially true as the use of reason to probe the secrets of nature was making breathtaking advances and being slowly popularized in the minds of common people by a few prolific writers of the time.

POPULARIZING REASON

During what is sometimes called "The High Enlightenment" in the mid-1700s, a number of French writers and philosophers produced a massive volume of highly controversial and influential works. These works expanded on many of the concepts in Locke's work, popularized many of the scientific advances of the previous century, and paved the way for a more complete transformation of our treatment of faith and governance. Primary among these writers and philosophers (and one of the first internationally acclaimed writers in history) was François-Marie Arouet, also known as *Voltaire*. Voltaire was famous for his wit and satire and was an extremely prolific writer, having produced tens of thousands of letters, books, and pamphlets in a variety of formats including novels, plays, poems, and polemics.

He was a staunch critic of the Church, aristocracy, and the State, as well as religious and class-based intolerance. He advocated for freedom of speech, freedom of religion, and separation of church and state (echoing many of the same sentiments as Locke before him). His criticisms and satires put him at constant risk of censorship and exile, eventually causing him to have to flee France for most of the end of his life. But they also served to popularize and normalize many controversial ideas that were otherwise considered taboo to speak about publicly in the eighteenth century. Things like the idea that every individual was born with the same right to liberty (whether peasant or king) or that we all start off in the same place (a "tabula rasa" or blank slate). Ideas like these were dangerous to the ruling elite as they encouraged greater independence and the questioning of authority.

Not only did he serve to popularize and normalize many of the concepts central to the revolutions of this period like freedom and equality, but he also popularized many of the extraordinary successes of early science. He published several scientific expositions like *Elements of the Philosophy of Newton (Éléments de la philosophie de Newton)* in 1738, which brought the groundbreaking science of Newton's *Principia* and *Opticks* to the masses. The experimental methods of *Opticks* and the mathematical methods of the *Principia* came together to form a new comprehensive model of natural philosophy, which later became known as "Newtonian science." Thanks to Voltaire's ability to make the sometimes arcane language of this new science understandable to the masses, Newtonian science found its way into mainstream thought and influenced many generations of scientists and philosophers to come in both Europe and the Americas.

Alongside Voltaire as one of the greatest literary and philosophical influences of this time is Jean-Jacques Rousseau, another significant writer and philosopher of the French High Enlightenment. Although Voltaire and Rousseau did not always see eye-to-eye, they agreed on many of the most pressing and controversial issues of the period, like inequality and the many abuses of power by Church and Crown. He openly criticized the Church, Crown, and aristocracy and was also under constant threat of arrest, exile, or harassment.

In his 1755 work *Discourse on Inequality*, Rousseau proposed that the moral iniquities of our time are not the result of man's nature, but the result of deviation *from* our state of nature into complex civil societies. He felt that as societies developed and we started to live in groups, people became more dependent on others to satisfy their needs, leading to new motivations and methods of self-preservation, including the desire for competition and dominance. The invention of property and the division of labor were not actually natural developments, but clever devices created by those with power (the rich) to preserve their standing. To prevent class divisions from escalating into all-out hostilities, they built political structures that would legitimize their advantage through laws, which would trick the poor into believing their freedoms were preserved when in fact, these same systems would ensure the inequities remained in place.

While in *Discourse on Inequality*, Rousseau identified many of the problems inherent to the establishment of political order beginning from our state of nature, in 1762, he published *The Social Contract* which presented a course of action for rectifying them. In this work, Rousseau contests the divine right of kings. This belief—which was dominant throughout Europe and much of the rest of the world for generations—was based on the idea that a monarch derived their authority to rule directly from a god or deity, and thus was not subject to any earthly laws or authorities.

As the use of reason to derive nature's principles made strides and was popularized in the minds of common people by writers like Voltaire, people's faith in humanity's innate faculties grew. Along with the greater independence of thought generally encouraged by Enlightenment-era thinking, this made it more feasible to contest the legitimacy of the divine right of kings, which had survived mainly because the idea itself went uncontested. In *The Social Contract*, Rousseau argues that the only way to preserve the individual liberties we are born with in our state of nature is to treat the whole group of people under a government as a collective "sovereign," in some ways as if they were one person. This is possible only if people enter into a social contract where they accept the authority of this government where they are themselves collectively represented. In this way, society becomes a reflection of the people's will and protects them against the abuses of the few. Rousseau's *The Social Contract* was one of the earliest calls for representative government and—along with Voltaire and Locke—greatly influenced the collective thinking of an increasingly global audience of readers on both sides of the Atlantic.

THE CONSOLIDATION OF WORLD KNOWLEDGE AND GLOBAL REVOLUTIONS

The influential literature of this time cannot be mentioned without also mentioning *the Encyclopedia*, also known as *a Systematic Dictionary of the Sciences, Arts, and Crafts (Encyclopédie, ou dictionnaire raisonné des sciences, des arts et des métiers)*. This was a project to bring together the combined knowledge of all the great advances in science and the arts up until that point and make it available to the masses. Alongside Jean le Rond d'Alembert as partial co-editor, *the Encyclopedia* was spearheaded by Denis Diderot, another French writer and philosopher of the High Enlightenment.

Between the years 1751 and 1772, Diderot worked for more than two decades to complete this project, writing as many as 7,000 articles himself. This was one of the first projects to include work from multiple authors (known as the Encyclopédistes), though many chose to remain anonymous given the controversial nature of the work. It was also the first comprehensive body of knowledge to address the mechanical arts, including descriptions of traditional craft tools and mechanical production processes, facilitating the creation of new and more efficient machines.

The largely secular tone of the *Encyclopédie*—often questioning many of the metaphysical or historical assertions of the Church, and critical of many of the abuses of Church and state—earned it a place on the Catholic Church's list of banned books and put its creators under threat by the authorities. Dovetailing with many of the sentiments echoed by other leading Enlightenment-era voices, the *Encyclopédie* made accessible to the general public knowledge that was otherwise inaccessible or controversial, contributing significantly to the revolutionary sentiment of the time. While these sentiments had a profound impact on the future of European political and cultural order, they were also being absorbed by an increasingly restless people across the Atlantic in the English colonies of the Americas.

The American Revolution, taking place roughly between the years 1765 and 1783, was a direct consequence of European Enlightenment-era thinking and contributed significantly to strengthening and reinforcing its core principles. American revolutionary thinkers like Thomas Jefferson, the principal author of the Declaration of Independence, were greatly influenced by the European intellectual environment of the time, with Jefferson calling Bacon, Locke, and Newton "the three greatest men that have ever lived." The principles of individual liberty, freedom, and equality, as well as the necessity to protect people from the abuses of Church, state, and aristocracy, were well represented in the values espoused by American revolutionary thinkers, and were eventually reflected in the United States Constitution.

The newly established America recognized the sovereignty of "the people" as the only source of legal authority and rejected aristocracy and hereditary political power. Concepts such as liberalism and republicanism, natural law, natural rights, and consent of the governed went against well-established ideological and political structures like absolute monarchy and the divine right of kings. This cemented a new political and cultural identity in the American mind, which influenced European politics in turn. The Americans' success at overturning British rule—seen as a great world power—gave renewed confidence that the same could be accomplished in other places around the world.

The French Revolution of 1789 is arguably one of the most significant events in world history. It often marks the end of the Enlightenment era and wider Early Modern Period. It was here that the ideological tide of the "Age of Reason" achieved its full material effect. "Reason" no longer belonged solely to the realm of natural philosophers but was now also a legitimate pathway to determining best principles in virtually every other field, including politics and governance. The new liberal principles put forward by thinkers like Locke, Voltaire, and Rousseau, alongside fundamental advances in science occurring in the previous century, permanently changed the ideological landscape in Europe and around the world. While the advances of the Scientific Revolution had a profound impact on our cosmological outlook, *the Enlightenment had a profound impact on our understanding of our own role within it*. These were not questions solely of the nature of physical objects in motion, but the nature of best principles as it relates to social organization and relationships.

The ideological shift caused by the new philosophies of the Enlightenment led to the system known as feudalism (dominant in the European landscape for many hundreds of years) to be all but abolished in France within a single year. The Church—which had owned nearly ten percent of all land prior to the revolution—was stripped of its property. The nobility were deprived of titles and estates, and the king's power was diminished to that of a figurehead. Equality won over in every field, from taxation to the ability to run for office to treatment under the law. The revolution's motto of liberty, equality, and fraternity (*liberté, égalité, fraternité*) was represented in nearly every level of this political reorganization—sometimes forcefully—and in a very short period of time. Aristocratic society was all but destroyed and replaced with the prominence of the "autonomous individual" with the freedom to act as they saw fit within the bounds of the law, equally applicable to all people, regardless of rank or station. The French Revolution had a profound impact on Europe and the rest of the world.

It is often the case when there are disagreements at the highest levels of governance that people begin to pay more attention to "philosophy," the reasoning behind why things are or should be a certain way. Rival leaders may need to put forward their ideological visions for governance, and people have to decide which philosophical approach they feel is most "right." It was the case in Europe at the end of the eighteenth century that the American and French Revolutions caused a widespread counterreaction. The victory of the "common person" over old elite structures like the aristocracy and the Crown made others in positions of power nervous.

There was a flurry of papers and publications arguing passionately both for and against the happenings in France and the Americas. As the new ideological principles of the political philosophers and the scientific advances of the natural philosophers made broad advances, the tide of revolutionary fervor spread far and wide across the world. A wave of revolutionary change spread throughout the Americas in the 1800s as a steady process of decolonization took place, and country after country gained their independence from European colonial rule.

THE BEGINNINGS OF THE SCIENTIFIC INSTITUTION

Meanwhile, the increasing body of knowledge coming from the newly established "scientific method" and the growing network of scientific academies and societies began to form the basics of an institution. Advances in physics and mathematics made new and incredible feats of engineering possible. Diderot's *Encyclopedia*, which included the first aggregated account of mechanical principles, along with new work from a growing network of novice philosophers and scientists worldwide, contributed to the development of new and more sophisticated machines. It was suddenly possible to mechanize a variety of production processes that allowed us to create products on a vastly larger scale.

This led to a period often called the "Industrial Revolution," occurring roughly between 1760 and 1840, which heavily emphasized the utility and power of machines. As our scientific knowledge exploded and our understanding of the mechanisms behind an increasing number of natural processes grew, the potential for new technologies to utilize those mechanisms grew in parallel. This led to new tools and innovations that would usher in an era where our understanding of mechanism and emphasis on quantitative and mechanical conceptions of nature dominated not only our industry and our engineering, but also our basic cosmological outlook.

The visible successes of this rationalistic, reasoned approach to cosmology, the conception of the world as a big machine that can be understood perfectly through careful examination of its causal laws, had a subversive effect on religious authority, especially over truth. While both science and religion put forward cosmologies, stories that account for the nature of things, science had the additional benefit of empirical evidence. In a science based on empiricism, you have ample physical evidence to demonstrate the validity of your claims. This stood in contrast to claims of religious truth, which relied on faith.

It was during the Enlightenment that we both established and moved past many first principles of science. We had the basis of geometry from Euclid's *Elements*, the classical mechanics of physics from Newton's *Principia*, and the establishment of basic methods and tools of measurement and quantification. We were now equipped with mathematical tools like calculus and geometry, physical tools like the telescope and the microscope, and classical mechanical concepts from Newton, Kepler, and Galileo. With a public increasingly interested in reason and natural knowledge and the proliferation of scientific societies and journals, the global environment was ripe for a rapid influx of change, innovation, and new knowledge. The successful application of reason to fields outside of natural philosophy like politics and ethics only served to compound this effect. This resulted in a gradual shift in our general worldview from that of religion to that of philosophy and science, and the use of reason rather than faith in the derivation of best principles.

It was around this time that our global outlook was transformed into the basic cosmological outlook we have today. Reason reigned supreme as the final arbiter of truth. The scientific method, based primarily on quantitative and empirical methods of study and observation, became the de facto method of inquiry into any field of academic study. As the volume of scientific data increased, it became more difficult to study nature as a whole. Serious academic study in any field required more background to catch up on the furthest extents of our thinking at the time. This led to a steady increase in specialization and "silos of knowledge," isolated domains of intellectual expertise cut off from a more holistic perspective on nature. We started to dive into the minutia, into the details of virtually every subject or area of knowledge, applying primarily quantitative and empirical methods to derive truth.

While significant developments in science and philosophy followed the Enlightenment which we will discuss in later chapters, it is the evolution of our social and cosmological outlook that is important to understand. Philosophy was the field that was born from the use of reason to derive natural principles. As primarily quantitative and empirical methods of study began to dominate, those inquiring into nature using these tools and methods began to see themselves as doing work different from the more general use of reason in the deduction of truth (or best principles) in other fields (or with other methods). What could be called more general philosophy. It was during and after the Enlightenment that the terms "science" and "philosophy" began to separate from each other, and the world's natural philosophers began to be known as "scientists."

PART II

THE SENSES

&

THE INTELLECT

CHAPTER 6

THE SUBJECT
AND THE OBJECT

By the end of the 19th century, many of the major facets of classical physics had fallen into place. The term "scientist"—invented in 1833 by English polymath William Whewell—began to replace the more common term "natural philosopher." The basic principles of fundamental physics like electromagnetism and thermodynamics were established by brilliant minds like James Maxwell and Michael Faraday. Concepts like the wave nature of light and the connection between heat and energy were discovered and expanded upon. Darwin laid out the foundations for modern genetics.

In the 19th century, our grasp of basic physical concepts led to an explosion of new technology. We invented the telegraph and the typewriter, the phonograph and the gramophone. The electric motor, induction motor, and internal combustion engines came on the scene. The first steam locomotives, automobiles, and tractors were invented. We created the first oil refinery, electric power plant, and the first light bulb. It was the century of dynamite and the machine gun. *All of this in the 19th century alone*, transforming virtually every area of human life for most of the world.

With such an incredible rate of change in such a short period of time, a change the human race had not seen with such rapidity in the whole of its recorded history, it was hard to argue with the success of "science," this offshoot of natural philosophy beginning to be recognized in its own right as a separate field, given its dedication to specific methods. Qualitative aspects of reality were deemphasized in favor of quantitative methodologies, mathematics, and physical relationships. As the study of nature (what was known as natural philosophy for nearly 2,000 years) slowly morphed into science, our endeavor to understand all things in the observable cosmos began to slowly narrow its focus.

THE EXCLUSION OF THE OBSERVER

There was an important change to our treatment of nature at this time which usually goes unnoticed. In the relationship between "the subject"—*the observer* doing the observations and picking up the sensory data—and "the object"—*the observed* thing whose qualities we are witnessing and measuring, there became an unspoken rule to exclude all aspects of reality pertaining to the observer (the subject).

The reasoning was roughly so. Since subjective reality is different for every observer, there is no way to study it and come out with objective results. It is inherent to the nature of subjectivity that it is different for everyone, therefore in order to truly understand nature, we have to keep everything to the realm of the purely objective. That way we can be sure that our observations are true for *everyone* and speak to the constant and predictable laws of our collective universe, not just to the realities that pertain to us individually.

Essentially, we performed a mental trick to facilitate our chosen methods of study. In our mental frameworks, "the subject" became "an object" so that we could treat with the whole cosmic order from the perspective of pure, physical mechanism. This facilitated greatly our ability to apply our methods to all data. Remember how the Ionians realized that nature followed objective processes, but their predecessors believed it was driven by "subjective entities" that controlled everything? Well when we trace the evolution of thought from Ionia until this period, we see that this belief evolved to such a degree that we began to treat nature as if there *were* no subjective entities. Everything was purely an objective process from which subjective entities somehow later emerged (though as something of an afterthought and unessential to its mechanics).

Famed Austrian-Irish physicist Erwin Schrödinger—responsible for "Schrödinger's cat" and "the Schrödinger equation"—briefly outlines the situation:

. . . Let me remind the reader about our main problem, which will be to find out the special and somewhat artificial features of present-day science that are supposed to originate from Greek philosophy. We shall submit and discuss two such features, namely the assumption that the world *can be understood*, and the simplifying provisional device of *excluding the person of the 'understander'* (the subject of cognizance) from the rational world-picture that is to be constructed. The *first* definitely originated from the three Ionian 'physiologoi', or from Thales, if you like. The *second*, the exclusion of the subject, has become an ingrained habit of old. It became inherent in any attempt to form a picture of the objective world such as the Ionians made. So little was one aware of the fact that this exclusion was a special device, that one tried to trace the subject within the material world-picture. ...We cannot trace the 'exclusion' as a definite step, decided upon consciously (which it probably never was).[1]

Using this trick in our mental frameworks allowed us to view the cosmos as a giant machine, leading to significant advancements in our understanding of physics and physical relationships. But it left a glaring blind spot with regard to subjective aspects of our reality that did not fit neatly into quantitative and physically-oriented models. It ensured that not only did we not understand these aspects with the same logical precision or integrity, but also that they were entirely absent from the cosmological model that arose from our accumulated scientific knowledge.

What's important to realize is that the exclusion of the "subject of cognizance," as Schrödinger puts it, did not come from observation or an attempt at recreating the world around us as it is. It was a "simplifying provisional device," as he puts it, meant to facilitate the study of certain phenomena. Realizing it did not follow from observation but was a simplification is important, because it makes clear that the exclusion of the subject from the study of nature meant, like any simplification, *excluding data*. It paints an incomplete picture of the whole.

This habit, given its success in helping us to understand the physical world, has become deeply ingrained. Like any habit which has persisted through successive generations, we have grown so used to it that we don't see it anymore. We've grown so accustomed to it that we tend to think it is the only way—or at least the only *correct* way—of doing things. As Einstein once said in his obituary for Ernst Mach (a philosopher influential in his own thinking):

Concepts which have been proved to be useful in ordering things easily acquire such an authority over us that we forget their human origin and accept them as invariable. Then they become "necessities of thought," "given a priori" etc. The path of scientific progress is then, by such errors, barred for a long time. It is therefore no useless game if we are practising to analyse current notions and to point out on what conditions their justification and usefulness depends, how they have grown especially from the data of experience. In this way their exaggerated authority is broken. They are removed, if they cannot properly legitimate themselves; corrected, if their correspondence to the given things was too negligently established; replaced by others, if a new system can be developed that we prefer for good reasons.[2]

The preference to exclude the subject from the study of natural phenomena was predicated on an assumption. Namely, that subjective aspects of reality do not follow discernible patterns in the same way physical processes do. Said another way, *we carry a root assumption that there are no objective processes (or natural laws) that can be discerned and understood as they relate to the conscious observer's relationship to their material environment.* We assume that because subjective experience is different for everyone, it is not subject to natural laws that can predict behavior. In essence, that cause-and-effect relationships only belong to the realm of physics, physical matter, and objects, but not to the realm of the subject, subjective experience, or to the subject-object relationship.

SUBJECTIVITY IN THE COSMIC ORDER

That subjective experience exists is irrefutable. What is not so clear is where to place it within the hierarchy of the cosmos. Where do we place the subject and all the observable aspects of their experience (this incorporeal yet tangible phenomenon), within the network of beliefs and concepts that form the basis of our thinking? The answer to this question is important, because depending on how we frame it, our entire cosmological outlook is affected.

If we believe that the subject arises from material objects, that the object is primary in the cosmos and that subjectivity emerges from it, then we will look to objects to try to understand it further (since they are its "cause"). Essentially, the core premise in this is that physics *causes* consciousness. What is important to realize is that this is not proven. *It is an assumption.* We do our best in science not to make assumptions, which makes it all the more ironic that we do so in this space, which is so important in the determination of the overall order of reality.

The question is, what if subject and object were actually two sides of one coin? What if the reality of the subject sits alongside the reality of the object as a matter of necessity? That the whole construct of reality could not work—could not be or exist—without it? If this were true, then looking more and more deeply at the object wouldn't help us, because the nature of the subject would be something requiring a different toolset to understand. If the subject does not emerge from the object, but is an equal opposite to it, then this changes the way we need to approach the question. How we place subjective reality within the hierarchy of our beliefs (our ontology) affects nearly every other area of human knowledge, *because it forms the foundation of the relationship between every observer and their material environment.* This has a bearing not only on our personal experiences, but also on our science.

To understand the situation better, we must make one more distinction. We must differentiate between the two primary tools with which we perceive the world around us (and upon which all science is based): *the senses and the intellect.*

THE SENSES AND THE INTELLECT

There is a fundamental duality to the nature of perception that encompasses two main areas. On the one hand we have *the senses,* what could be called "sense-data" in the form of information we receive through the senses. And on the other hand we have *the mind* or *the intellect,* what is sometimes called "mentation" including mental conceptualization and imagination, as well as reason for the ordered structuring of ideas and phenomena. Even in early Greek times there was a recognition that these two areas were fundamental in the interpretation of reality.

It is no secret that the senses can sometimes deceive us, which is to say, present us with a version of reality that is inaccurate or distorted. Consider the visual perception of objects through water. Light is often distorted, giving us the perception that something may be bigger or smaller than it actually is, or that its shape is different. In a situation like this, we may require more information to accurately assess the data we are receiving through the senses and form an accurate perception of things.

The intellect may seem like a cold calculus, that its adherence to rationality or logic make it immune to this type of distortion, but even it can be deceived. The understanding reached through logic and reason is dependent on the accuracy of the associations drawn between the various phenomena it identifies. Sometimes we make mistakes in logic or inferences based on incomplete information. These errors in reasoning can be at a surface level or *they can be very deep in our logic structure.* Statements close to the root of our logic tree have bearing on entire branches of reasoning, and errors can put conclusions in far-reaching areas of our logical frameworks on shaky ground.

The scientific method, established during the sixteenth century with our brilliant post-Renaissance minds, is based on both the senses and the intellect. The aspect associated with the senses we call empiricism. The basic premise behind the use of empiricism is that *only things we can experience first-hand can we confirm to have validity in "the real world."* We "witness" phenomena with our senses, like with sight or hearing for example, and make record of our observations, doing our best to be as neutral as possible, careful not to make personal opinions about things, but to observe things "as they are." This empiricism exists on both sides of the scientific process, as the method by which we collect our data to analyze, and how we ultimately test our hypotheses to verify their accuracy.

The aspect of the scientific method associated with the intellect could be called "rationality" or "reason." *Reason is a form of pattern recognition that occurs through the linear ordination of cause-and-effect relationships.* To put it simply, the intellect takes the data we receive through the senses and creates *order* out of them. Through the intellect we can understand nature in a less tactile and sensory way, and in a more neutral, technical, and objective way.

Even our reasoning takes on one of two forms—*inductive* and *deductive*—each belonging more to the senses or to the intellect. On the side of the senses we have inductive reasoning. With this type of reasoning we take data about individual phenomena we collect via the senses and piece them together to figure out broader or more universal patterns or principles. In other words, we begin with the specific and move to the more general or universal. Opposite to the inductive reasoning emphasized by the senses, we have deductive reasoning emphasized by the intellect. Here we begin with general or universal principles and apply them to individual phenomena, leading from general conclusions to more specific ones.

Through the combination of these two modes of perception—the sensory and the rational, or, the sensory and the mental—we analyze all the various phenomena we witness and experience in the world. It is through their sum total that the material world is comprehensible to us. We return to Einstein who lucidly outlines this relationship:

> In speaking here concerning "comprehensibility," the expression is used in its most modest sense. It implies: the production of some sort of order among sense impressions, this order being produced by the creation of general concepts, relations between these concepts, and by relations between the concepts and sense experience, these relations being determined in any possible manner. It is in this sense that the world of our sense experiences is comprehensible. The fact that it is comprehensible is a miracle.[3]

You may notice that he used two terms here that may not be familiar to you. In academic literature, there are sometimes two terms used when referring to sensory relationships: sense-impressions and sense-experiences. It may sound like splitting hairs, but the distinction is critical to understanding what sets the senses apart.

Sense-*impressions* refer to a process that occurs from "the outside-in," so to speak. Physical stimuli like a lightwave or a vibration make an impression on one of the physical sensory organs, like the eye or the inner ear, for example. These impressions can be measured in the same way any other physical thing can be measured, given that they are impressed upon the organ from the outside. This could include the frequency of a lightwave hitting the eye or a change to the organ as a reaction to the impression.

Sense-*experiences*, on the other hand, are speaking of this same occurrence from the opposite side, from the inside-out. An "experience" speaks to the subjective interpretation of that impression, regardless of its physical attributes. This is subject to other factors such as one's beliefs (psychological framework) or it can be distorted because of something occurring in the body (like an ailment or illness). These factors are distinct from those that are purely physical. Either way, both sense-impressions and sense-experiences pertain to a *subject*, what could be termed an "observer" or experiencer. This is placed in opposition to the reality of an "object," a material constituent in a particular physical formation.

What can be gleaned from all of this and is important to understand, is that *sensory data (by virtue of sensory experience) does not always pertain to that which is physically visible or materially tangible.* This simple point opens up the inquiry into elements of nature beyond those which can be seen and measured with material instruments.

THE BRIDGE

While experiences may not be physically measurable, they are tangible from the perspective of the observer. The core question is this: Between the senses and the intellect, what is the totality of data that is observable to us and needs to be incorporated and accounted for in order to formulate a cosmology which speaks to everything? We must determine how it is that all of the data fits together as a whole, and create a model that can work for all of it. If subjective and objective elements seem to be at odds with each other, rather than throwing out data, it is better to ask, how can both sides of this equation fit together within the same ontological model or framework? Where is the bridge between them?

What is required is a new way of looking at things. One that doesn't leave anything out. Nothing we have learned about physics, and none of the various data we have about our subjective reality. A science that speaks to the whole is capable of forming a logically coherent bridge between inner and outer elements of the observable cosmos.

It was not until the early 20th century that science was again confronted with questions regarding the nature of the subject, in relation to the objective measures and physical relationships it was used to looking at. It appeared that our attempt to exclude the observer was nearing the limits of its usefulness in the examination of physical relationships. It was around this time that the rules of classical physics, which had held up for hundreds of years, found their first serious challenge. These came primarily from two groundbreaking theories which revolutionized our view of the world: the theory of relativity and quantum mechanics.

CHAPTER 7

THE OBSERVER

The fundamental conclusions of classical physics held up for quite a long time! The idea of a static physical world independent of the observer, behaving according to strict causal relationships (what is sometimes called "causality") and determined by the local interaction between objects (sometimes called "locality") was the conceptual model that catapulted natural philosophy into the realm we call science. We were able to determine so many basic physical principles: laws of motion and gravity, thermodynamics, electromagnetism, the basics of chemistry and biology. Our understanding of "mechanism," at least as it relates to physical relationships, exploded. Consequently, so too did our capacity to build machines which went on to change the world.

It was not until the late 19th to early 20th century that things began to change. One of the most important discoveries of the late 19th century came from Scottish physicist James Maxwell. Famously called "The Maxwell Equations," he put forward four equations that established the basics of electromagnetism. These equations helped us understand that there was a fundamental relationship between electricity and magnetism (some of which had already been established earlier that century by Michael Faraday), and that electric charges and currents create electric and magnetic fields. Furthermore, Maxwell used these equations to suggest that "light" was really a form of electromagnetic energy. He stated:

> We have strong reason to conclude that light itself (including radiant heat and other radiation, if any) is an electromagnetic disturbance in the form of waves propagated through the electro-magnetic field according to electro-magnetic laws.[1]

This effectively unified the fields of electricity, magnetism, and optics. Further, the discussion of whether light was most fundamentally *a wave* (sometimes called an "undulation") or *a particle* (sometimes called a "corpuscle") was sent into its final phase. It appeared from Maxwell's equations that light behaved like a type of electromagnetic radiation that moved through space as *a wave*.

This conversation had been going on for quite some time. Sir Isaac Newton, for example, had put forward a particle-based theory of light in his famous work *Opticks*. Some of his contemporaries (and Descartes before him) put forward theories in support of a wave theory. It appeared that Maxwell's equations had put an end to that debate, as the electromagnetic theory of light was confirmed by another famed scientist, Heinrich Hertz, soon after, eventually leading to the development of new technologies like radio, television, and wireless communication. But it was not long after this, at the dawn of the 20th century, that the particle-based theory of light was again revived by two leading scientists of the time—Max Planck and Albert Einstein.

It was also at this time that the basic principles of classical physics (a.k.a. classical mechanics), which had enjoyed a position of unchallenged authority in the scientific community for nearly 400 years, were challenged for the first time. It appeared that the rules of classical mechanics held up for most objects we could witness in the world, but when our tools of observation grew more sophisticated, and we began to analyze events that were at scales vastly smaller or larger than we were accustomed to, our classical model began to fall apart. It was then that two new theories came into the picture which revolutionized our understanding of the physical world for the second time since Newton's *Principia*: the theory of relativity and quantum mechanics.

A REVIEW OF OUR PICTURE OF THE PHYSICAL WORLD

In order to make clear some of the advances in our understanding of the physical world that were brought on by quantum mechanics and the theory of relativity, it will help to briefly outline what our concept of the physical world was up until that point.

Two of the most important of these physical concepts are called "locality" and "causality" (or what is sometimes referred to as "causal determinism"). The concept of locality simply states that all physical interactions occur "locally," or within a given space. In other words, objects cannot interact with each other at a distance. If a system is disturbed or changed in some way, it is because it has come into *direct physical contact* with something else. The change that occurs in this system because of that interaction may then, in turn, cause a change in another system that is within *its* local environment, and so on and so forth like a physical chain of events. This creates a clean causal chain that always occurs by way of direct physical contact between two or more waves, particles, forces, or systems. This is the concept of locality.

Loosely speaking, the concept of causal determinism or "causality" is as follows. The universe is governed by laws, and these laws can be discerned and understood. All objects in the universe obey these laws, therefore if we understand all the laws that govern these objects, we should be able to predict every action and reaction all the way from the present moment until the end of time. In other words, what will happen later is *already determined* by the causal chain of events that follows from these natural laws. Add into this that all "events" occur locally, and you have a picture of the world where the physical interaction of one object on another, and all of the laws governing those interactions can determine everything that has ever happened and that will ever happen in the universe. This is the "determinism" part of causal determinism. The future and the past are essentially already determined by the natural laws that govern the interaction between objects, so nothing occurs at random. Together, locality, causality, and determinism formed our basic interpretation of the nature of reality for many hundreds of years.

There are two more concepts I will introduce to complete this picture, what is sometimes called "strong objectivity" and "epiphenomenalism." Please don't get too bogged down by the vocabulary, I know it sounds like a lot. We will focus on the *meaning* of the concepts, rather than the words themselves, so bear with me. Strong objectivity, which sometimes goes by other names, basically states that the physical world exists independently of whether we observe it or not. There is an "objective" real world that keeps on existing irrespective of any "subject" or consciousness. Essentially, reality is defined by physical processes that exist irrespective of consciousness. This is more or less strong objectivity, the "objectivity" part being the exclusion of the subject, and the "strong" part meaning that it is absolute.

Epiphenomenalism is the idea that consciousness, the conscious observer who is *doing* this science, is an "epi" or "sub" phenomenon of physics. Consciousness and the consciously experiencing observer are born from and produced by physical objects. It is an "epi" (meaning stemming from or born from) phenomenon.

Together, these four concepts helped to define our perception of reality or the "real world," for many years. To summarize, the idea was that at any given moment, there are a set of physical objects interacting locally (locality), which are governed by natural laws that can perfectly predict their behavior (causality), all existing independently of consciousness or any subjective observer (strong objectivity), and which at some point came together to create "the observer" (epiphenomenalism). This created the picture of a static physical world with perfectly predictable behaviors and perfect chains of cause and effect determined by discernible natural laws, which eventually "caused" consciousness. Please keep these concepts in mind as overarching themes. We will dive more into the specifics now, and return to the higher, more general level afterwards.

THE ATOM

It was also around this time that we developed a clearer picture of the basic components that come together to form our physical world. At the most basic level, we have "the atom." The idea of an atom, a fundamental physical unit from which all other physical materials are constructed, was actually first conceived of around the 5th century BC, at the same time as our original philosophers.

It was a Greek philosopher named Democritus who theorized that there must be something that exists at the limit of divisibility. In other words, if you kept cutting something in half or taking it apart, eventually you would arrive at *a fundamental substance*, and could not "cut" or reduce it further. In Greek, the prefix "a" meaning "not" and "tomos" meaning "cut," came together to form "atom," which essentially meant "uncuttable" or indivisible. The implication of this was that there was no physical substance this could be further reduced to, and all other things were constructed from it.

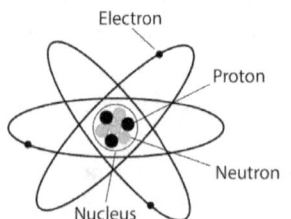

It was not until 1897 that British physicist J.J. Thompson discovered that there was actually something smaller than an atom. Thompson is credited with discovery of *the electron*, a particle contained *inside* the atom, significantly smaller in size, and with a negative electrical charge. Less than twenty years later, in 1911, New Zealand physicist Ernest Rutherford discovered that at the center of every atom was also a "nucleus," a conglomeration of two other types of fundamental particles. There was a positively charged *proton* (coming from the Greek prôtos (πρῶτος) meaning "first") and a neutral or chargeless *neutron* of roughly the same size and weight, that were "glued" together at the center of every atom in what was called its "nucleus." The electrons, thousands of times smaller in size and weight, would orbit this nucleus at unbelievable speeds.

The number of protons and neutrons in the nucleus of the atom would then go on to determine some of its other characteristics, like weight and charge, which would ultimately determine what "element" it was (e.g., hydrogen, oxygen, gold, silver, etc.). These fundamental elements, which were essentially atoms of different weights and with different numbers of protons in their nucleus, were organized into what is called the "Periodic Table of the Elements." Basically just a catalog of all of the naturally occurring (and some artificially created) basic atomic configurations.

To complete this picture, these "atoms" would then join together to form *molecules*, stable configurations of multiple atoms that came together to form a new physical substance. A molecule of water, for example, is made up of 2 hydrogen atoms and 1 oxygen atom. These three atoms come together in such a way that the electrons share the whole molecular structure, encircling not only one atom, but all three. The sharing of space around all three atoms helps to keep the molecule stable, given the counterbalance between their negative electrical charge and the positive charge of the nucleus.

In order to picture the size of these atoms, consider that in a single drop of water, there are approximately 5 sextillion atoms (that's 5 followed by 21 zeros). They are small! This was more or less our concept of the basic formation of physical objects up until this point.

LIGHT

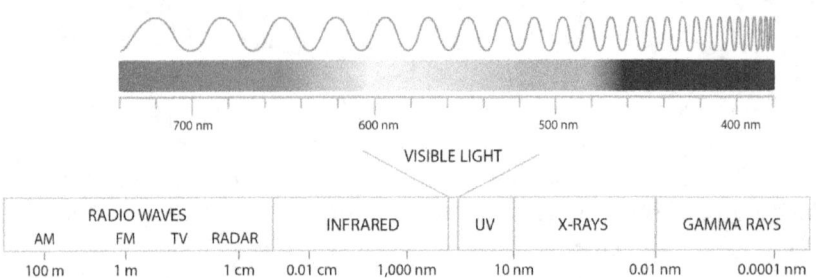

RADIO WAVES				INFRARED	UV	X-RAYS	GAMMA RAYS
AM	FM	TV	RADAR				
100 m	1 m		1 cm	0.01 cm	1,000 nm	10 nm	0.01 nm	0.0001 nm

For the average person, when we think of light, we are typically thinking of *visible* light. In reality, light is a more general phenomenon that can express itself in a variety of ways, both below and above the range of light that is visible to our eyes. We had been studying light for almost two thousand years and made significant advancements with work like Newton's *Opticks*, but it was not until Maxwell's equations and discovering the connection between electricity and magnetism, that we were finally able to identify light as a form of electromagnetic radiation.

The range of light that is visible to our eyes makes up a spectrum. At the bottom of this spectrum is red, followed by orange, yellow, green, blue, indigo, and violet—the colors of the rainbow. The reason these colors follow a particular order is because they correspond with what are called *wavelengths* and *frequencies*. A wavelength is the physical distance (the "length") between two waves. Frequency refers to how often the wave occurs within a particular period of time. Therefore light with a shorter wavelength will appear more often within a given amount of time, and light with longer wavelengths will appear less often.

Light at the low end of the electromagnetic spectrum (like red) has a longer wavelength and a lower frequency, while light at the high end of the spectrum (like violet) has a shorter wavelength and a higher frequency. Light with a higher frequency carries more energy.

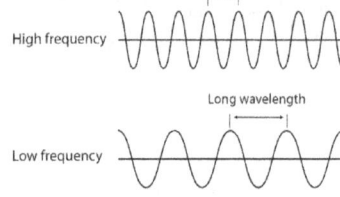

If you shine pure white light (like sunlight) through a prism, you will see it refracted into all the colors of the rainbow, which is basically a representation of all of the colors in the spectrum of visible light, in order from lowest to highest frequency.

In the 19th and 20th centuries, we also discovered that there was light both below and above the spectrum that is visible to our eyes. Below red light we discovered what is called "infrared" light, with the prefix "infra" meaning below. This is light with frequencies even lower than red. Some examples of this include radio waves (like we use in radio broadcasts), microwaves (like we use to warm up our food), and radar. Infrared light is not visible to the human eye, but warm objects produce a lot of it (including the human body). We usually experience this type of light in the form of heat, including a large part of the energy produced by the sun.

Above violet light we have what is called "ultraviolet" light, with the prefix "ultra" meaning "beyond." Some examples of ultraviolet light include so-called "black" lights, X-ray imaging (to see inside the human body), and gamma rays (which we use to kill certain types of cancer cells, and which are sometimes produced by large stellar bodies).

Light is sometimes called "electromagnetic radiation" because it is essentially electromagnetic energy "radiating" through space. The entire range of possible wavelengths and frequencies of light or electromagnetic radiation is what makes up "the electromagnetic spectrum."

SUMMARY

In summary, our general concept of the physical world consisted of impossibly small atoms with three fundamental particles making them up: a positively charged proton and neutrally charged neutron glued together in a "nucleus," and negatively charged electrons, much smaller in size, orbiting at unbelievable speeds. The number of protons in the atom would change what basic "element" the physical substance was, like hydrogen, oxygen, or nitrogen. These "elements" then came together to form molecules, stable configurations of multiple atoms, like a water molecule for example, which consists of two hydrogen atoms and one oxygen atom.

These atoms and molecules behave according to strict causal laws, and the interactions between them produce predictable chains of cause and effect that allow us to know both the outcomes of these interactions (going forward in time) and what led to them (moving backwards in time). Subjective elements of reality are excluded from this objective and "exact" science of physical objects and their interactions. It is assumed that consciousness, like every other thing in the universe, is ultimately a product of these same physical constituents and their interactions.

None of these concepts posed any serious challenge to the conventional view of physics begun with Newton's *Principia*, and continued for many hundreds of years until the turn of the 20th century. It was then, at that pivotal moment, that our basic view of the physical world changed yet again. German physicist Walter Heitler fluently summarizes this development:

> The beginning of the twentieth century is the milestone of a marked change in the direction of scientific thought as far as the science of the inanimate world, i.e., physics, is concerned. What is now termed classical physics is an unbroken line of continuous development lasting roughly 300 years, which began in earnest with Galileo, Newton and others and culminated in the completion of analytical dynamics, Maxwell's theory of the electromagnetic field and the inclusion of optics as a consequence of the latter by Hertz. If we also include Boltzmann's statistical interpretation of the second law of thermodynamics, the list of what constitutes the major parts of classic physics will be fairly complete.
>
> The logical structure of all these theories is roughly as follows: The happenings in the outside world (always confined to dead nature) follow a strictly causal development, governed by strict laws in space and time. The space and time in which these events occur is the absolute space-time of Newton and identical with what we are used to in daily life. The term "outside world" presupposes a sharp distinction between an "objective" outside reality, of which we have knowledge ultimately through sense perception, but which is completely independent of us, and "us," the onlookers, and ultimately those who think about the results of our observations. The term "causal development" is meant in the following, rather narrow, sense: Given at any time the complete knowledge of the state of a physical object (which may be a mechanical system, an electromagnetic field, etc.), the future development of the object (or for that matter, its previous development until it has reached the state in question) follows then with mathematical certainty from the laws of nature, and is exactly predictable.[2]

It is now that we dive into some of the deeper implications of the changes to our general worldview during this period. Though an effort has been made to render this accessible to the reader, if one prefers to skip the more technical explanations and get to the heart of the matter, it is recommended to skip forward to the final section entitled "The Observer." Otherwise, we dive into the fascinating developments of quantum mechanics and the dual wave-particle nature of light.

QUANTUM MECHANICS AND THE DUALITY OF MATTER

One of the most significant changes to the way we viewed the physical world starting in the early 20th century was our treatment of matter at both very large and very small scales. On the small side, we had begun to develop a picture of what was happening with the smallest unit of matter, the atom, and the particles that made it up: protons, neutrons, and electrons. The discovery of these subatomic particles opened up a whole new universe of study into the physics of small objects, especially electrons, whose size was even a fraction of that of protons and neutrons. Arguably one of the most important of these discoveries was the dual "wave-particle" nature of matter.

It is one of the most basic properties of all substances that there is a sort of taking in (absorption) and giving out (emission) of energy. This emission and absorption is sort of like breathing. Nothing purely takes energy in or purely gives energy out, but some substances can come close. When a substance absorbs almost everything it encounters, all of the wavelengths it comes into contact with, we call it a "black body." This is because it is taking in almost all electromagnetic radiation or "light" it comes into contact with, without reflecting anything back. Hence it appears "black." It is the ultimate absorber.

While on the one hand, all substances in nature are seeking order and equilibrium (based on forces like gravity or electromagnetism), on the other hand, they are also moving toward disorder. This natural movement toward disorder or chaos is what we call "entropy." The near total absorption of electromagnetic radiation by black bodies makes the light that they do emit, what is called "black body radiation," ideally suited to the study of entropy, since it is not mixed with reflected light.

In the year 1900, German theoretical physicist Max Planck made a discovery related to entropy that would go on to change physics yet again. What Planck's mathematics revealed was that the light emitted by black bodies appeared to have very specific wavelengths. Rather than carrying values at *any* point along the electromagnetic spectrum, they appeared to come only in *fixed intervals* proportional to a specific mathematical value, now called the "Planck constant." Essentially, light being emitted by these substances could not take on *any* value along the electromagnetic spectrum, but instead would only come in specific intervals that were proportional to the Planck constant. This meant that the energy was "quantized."

The Planck constant was a sort of minimum amount of energy that was required for the energy to be emitted, what was then called a "quantum" of electromagnetic energy (and later a "photon"). The term "quantum" refers to the smallest, indivisible unit or chunk of something. In nature (like in the case of light), this occurs because the phenomenon only happens in set, predictable amounts, with the smallest interval between them being a "quantum." In this way, we call the energy "quantized," meaning that it occurs in "chunks" rather than at any point along a spectrum. This is a bit like saying that when you look at the number line, you can only choose whole numbers but never fractions (1, 2, and 3, but never 1.2 or 2.3). It is this "quantization" of energy—that it does not occur in a continuous way like a clock hand spinning smoothly around a clock, but in a dis-continuous way, like the clock "ticking" from one second to the next in specific, set intervals— from which "quantum" mechanics gets its name.

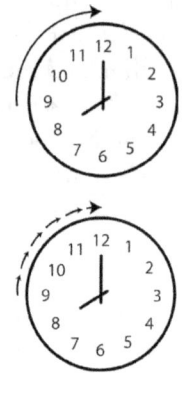

As we discussed earlier, Maxwell's equations had seemed to put an end to the discussion about whether light was a particle or a wave. Light was electromagnetic radiation traveling through space, and was therefore a wave. But this didn't fit with Planck's discovery. A wave should not come in "discrete packets of energy" or "quanta," but should instead be able to express itself at any point along a spectrum. Discrete units like this were more a sign of particle behavior, not waves, and so at least at first, Planck was reluctant to accept the implications of his own results. He thought the scientific community might disprove him, but quite the opposite happened.

Several years later, renowned German theoretical physicist Albert Einstein was able to connect Planck's idea with what is called the "photoelectric effect." The photoelectric effect says that when light is shone on a metal, electrons get dislodged from it. What was interesting about the photoelectric effect was that it didn't matter the intensity of the light or for how long you shone it on the metal. All that mattered in producing the effect was the *frequency* of the light. A high-intensity light at a low frequency would not dislodge any electrons from the metal, but a low-intensity light at a high frequency would. Therefore, this effect of dislodging electrons from the metal seemed to be connected to the frequency of the light that was hitting it.

Following from Planck's experimental conclusions, Einstein theorized that light energy was actually being carried in discrete, quantized packets. The energy in these "packets of light" was equal to the light's frequency, multiplied by a universal constant (later called the Planck constant). This "light packet," or discrete unit of light, was later dubbed a "photon." What Einstein discovered was that only a photon above a certain frequency had the energy required to eject an electron and create the effect. Einstein's theory—while agreeing nicely with experiment—was resisted for some time, because it seemed to go against the wave theory of light from Maxwell's equations.

What was eventually discovered was that when an atomic system changes in energy, the electrons that orbit its nucleus "jump" from one track or shell to another. Rather than the electron being able to be at any distance from the nucleus, it was only permitted to be within certain fixed distances we could call "tracks" or "shells." The atom would then emit or absorb energy according to the energetic difference between these orbital shells. Depending on the amount of energy introduced into the atom, an electron could jump from "shell 1" to "shell 5," for example, or from "shell 3" to "shell 4", but it would never appear at shell 3.5. The amount of energy required to take it from one shell to another was fixed, and could therefore be exactly predicted.

These packets of energy that were being either absorbed into or emitted out of the atom were photons, fixed packets of electromagnetic radiation or "light." These photons are "quantized," meaning they come in discrete, countable amounts rather than being at any point along a spectrum. The discrete amounts that these packets of light come in then correspond to the energy differentials between the different electron orbits. The startling discovery of this (which incidentally earned Einstein the Nobel Prize in 1921) seemed to suggest that in addition to light being able to propagate through space as a wave, it could *also*, at the same time, behave like a particle.

In *Physics and Philosophy* from German theoretical physicist Werner Heisenberg—one of the main pioneers of quantum mechanics—this is summed up nicely:

> In what became known as the old quantum theory, originating with Niels Bohr in 1913, atoms were pictured as little solar systems. Electrons orbited the small, massive nucleus strictly according to the principles of Newtonian mechanics. The quantum principle

came into this model with the additional restriction that only certain orbits, out of the infinite range possible, were in fact permitted. When an electron jumped between orbits, the atom either took in or gave out a quantum of electromagnetic energy—later dubbed a photon—corresponding to the energy difference between the orbits. This mechanism explains why atoms, as had been known for decades, had characteristic spectroscopic signatures, emitting and absorbing light only at certain fixed frequencies.[3]

These experimental and theoretical results eventually led to the realization that light was not really a particle *or* a wave. It was *both*.

QUEER QUANTUM BEHAVIOR - SUPERPOSITION AND ENTANGLEMENT

Despite the counterintuitive notion that light was neither strictly a particle nor a wave, but rather both (or neither), it was not the only idea to come from quantum mechanics that challenged our conventional notions of reality. What has become one of the most famous quantum experiments of our time is the "double-slit experiment." In this experiment, light is fired at a metal plate through two narrow slits. The light passing through these slits hits a screen behind them which records its impact.

When *waves* pass through these slits, they interfere with each other on the way to the recording wall and produce an interference pattern, showing brighter and darker bands of light. This is similar to what happens to two waves interacting with each other on the surface of a pond. Where they are very close, they sometimes cancel each other out, and at others reinforce each other. Particles, on the other hand, do not create interference patterns like this because they act as individual entities, passing through the slits one at a time. They make a different pattern. This was an easy way to tell whether it was a particle or a wave passing through the slits.

We already know that light can be both a wave and a particle depending on different conditions, so what happens when we pass it through the slits? Will it show its wave qualities or its particle ones? Astoundingly, it exhibits both depending on the way we choose to measure it. When we don't measure which slit the light goes through, it produces an interference pattern, demonstrating its wave-like nature. However, if we place a detector at the slits to determine which one the light passes through, it starts behaving like a particle, showing no interference pattern. In essence, the very act of observation causes the light to "choose" a behavior. This is not merely a limitation of our equipment, *it appears to be a fundamental principle of nature*. It's as if the light itself "knows" it's being observed and changes its behavior accordingly!

This is an example of a concept called "superposition," a fundamental principle in quantum mechanics. While in superposition, particles exist in multiple states at once—both particle and wave—until they are measured. Only upon measurement does the particle "collapse" out of superposition and "choose" a state.

Now what happens when we pass an electron (rather than light) through the slits? Electrons are particles right? So we would expect them to produce the pattern particles make in our example above. Yet after conducting this experiment, we learned that they produce the interference pattern we would expect from waves! The interference pattern produced by electrons is identical to one created by light of a particular wavelength. We discovered that in the same way *light* can exhibit particle-like and wave-like properties, *matter can too*. This gave rise to the term "matter-waves," indicating that while matter may appear to be solid and have a specific position, it can also behave as a wave. The "quantum objects" from which all macroscopic matter are made, *all* exhibit wave-particle duality. Returning to Heisenberg's book:

> Is an electron a wave or a particle? The answer, as Heisenberg insists in these essays, is that the words "wave" and "particle" are formalized in classical mechanics by derivation from our everyday experience, and by definition are mutually exclusive. A wave can't be a particle and a particle can't be a wave. A quantum object, in itself, is neither one thing nor the other. If you decide to measure a wave-like property (wavelength, for instance, in a diffraction or interference experiment), the thing you are observing will look like a wave. Measure a particle property (position or velocity), on the other hand, and you will see particle-like behavior.[4]

As further confirmation of this, imagine now that only one particle is passed through the slits at a time. Surely this would produce a particle-like pattern, given that each one moves through the slit alone and is recorded on the back wall one at a time. Yet it turns out that after passing multiple particles through the slits one by one, they produce a wave-like interference pattern as if each particle is interfering with itself!

One way or the other, our takeaway is this: *the act of observation af-*

fects objects at the quantum level. The way we choose to measure quantum systems affects what qualities they display to us. If we choose to measure wave-like properties like wavelength or frequency, we find wave-like properties. If we choose to measure particle-like qualities like position or velocity, we find particle-like properties. Measuring one is at the expense of any certain measurement of the other. When you measure particle-like properties, you "cause" the object to "be" a particle. Measure wave-like properties and you "cause" the object to "be" a wave. The implication of this is that it may in fact be impossible, when considering this quantum property, to have any totally objective measure of an object in nature. Schrödinger speaks of the implications of this:

> I say this interpretation suggests itself: that there *is* a fully determined physical object in existence, but I can never know all about it. However, this would be a complete misunderstanding of what Bohr and Heisenberg and those who follow them actually mean. They mean that the object has no existence independent of the observing subject. They mean that recent discoveries in physics have pushed forward to the mysterious boundary between the *subject* and the *object*, which thereby has turned out not to be a sharp boundary at all. We are to understand that we never observe an object without its being modified or tinged by our own activity in observing it. We are to understand that under the impact of our refined methods of observation, and of thinking about the results of our experiments, that mysterious boundary between the subject and the object *has broken down.*[5]

The duality of matter and the role of the observer have thrown our classical ideas of reality into disarray. The double-slit experiment suggests that the state of a quantum system is not determined until it is measured. This has led to many interpretations and debates about the nature of reality itself, including thoughts on determinism, causality, and the role of consciousness in the universe. It's important to note that the interpretation presented here is the Copenhagen Interpretation (proposed by Niels Bohr and Werner Heisenberg in the 1920s), which is the most widely taught and discussed in the scientific community today.

The implication of these results extends beyond the lab. Regardless of how comfortable we might be with the idea, the choice of experimental setup—made by a "subject"—directly impacts its results. Saying that it is "the equipment" and not the observer is an attempt to sidestep the unsettling notion that subjectivity may be deeply woven into the fabric of reality. It's not merely a semantic game but a deep ontological issue. Whether it is through direct observation or by way of a measuring apparatus, the subject is inextricably involved in shaping what we call "reality" at the physical level.

As staggering as these revelations are, quantum mechanics does not stop at challenging our understanding of particles and waves. When it comes to the queer behavior of the physical world at the quantum level, there is one other phenomenon that stands head and shoulders above the rest, as a hallmark of the limitations of our classical model in interpreting reality—*quantum entanglement*. This phenomenon, first highlighted in a thought experiment by Einstein, Boris Podolsky, and Nathan Rosen[6], shows that two or more particles can become correlated in such a way that the state of one immediately influences the state of the other, no matter how far apart they are. This "spooky action at a distance," as Einstein once called it, has been experimentally confirmed[7] and stands as one of the most mystifying aspects of quantum mechanics.

Quantum entanglement pushes the boundary of what we understand about locality and causality. If two entangled particles are light-years apart and the state of one is changed, the state of its entangled partner will change *instantaneously*. This instantaneous action seems to defy our conventional understanding of space and time, and has led to various interpretations and heated debates about the fabric of reality itself.

Just as the double-slit experiment calls into question the classical concepts of particles and waves, quantum entanglement raises profound questions about the nature of interconnectedness and causality in our universe. It brings to the fore yet another layer of intricacy, suggesting that our classical intuitions about the separability of objects and events might be fundamentally flawed. In both cases, the implications reach beyond the laboratory, casting a shadow of uncertainty over our traditional views of reality.

UNCERTAINTY AND THE PROBABILISTIC NATURE OF PHYSICS

So far we've learned about the duality of matter. That at both macroscopic and microscopic levels, physical matter has a dualistic property in which it can behave particle-like (concentrated into a point) or wave-like (dispersed throughout space). This duality is a fundamental aspect of physical reality, especially visible at the quantum level. Further, we've learned about the role of the observer in collapsing quantum objects from their "superposition" into more definite locations in space and time. Finally, we've learned about the fundamental entanglement that can occur between particles that allows them to communicate with each other instantaneously, irrespective of distance. All of these ideas challenged our conventional notions of physics and revolutionized our worldview. Yet as baffling as these phenomena are, they point at yet another fundamental principle that lives at the heart of the quantum world—*uncertainty*.

Our classical model of physics was based on a worldview where there were definite objects with clearly defined boundaries and strict causal laws that could predict their behavior exactly. In this model there was no room for uncertainty, only aspects of reality we had not discovered yet. Precision and exactitude were clear qualities of this science. Yet the quantum world did not follow the same set of rules. *It turned out that there were aspects of quantum objects we couldn't know about all at once with certainty.*

This "uncertainty principle" was named after its proponent, Werner Heisenberg. The Heisenberg Uncertainty Principle states that it's impossible to accurately know both where a particle is (its position) and how fast it's moving (its momentum) at the same time. The more closely you track its position, the less you can know about its speed. The more precisely you measure its speed, the more unclear becomes its position. In essence, the more you know about one thing, the less you know about the other. This was not because of some deficiency in our measuring apparatus or fault in our mathematics. Instead, *it was a fundamental limitation of the universe itself.*

This meant that, practically speaking, we couldn't know everything about a particle. It is this uncertainty that leads to the probabilistic nature of quantum mechanics. Instead of saying exactly where the electron is and how fast it's moving, we can only talk about the probabilities of finding it in different places with different speeds. Austrian-born American physicist Fritjof Capra comments:

> At the subatomic level, matter does not exist with certainty at definite places, but rather shows 'tendencies to exist', and atomic events do not occur with certainty at definite times and in definite ways, but rather show 'tendencies to occur'. In the formalism of quantum theory, these tendencies are expressed as probabilities and are associated with mathematical quantities which take the form of waves. This

is why particles can be waves at the same time. They are not 'real' three-dimensional waves like sound or water waves. They are 'probability waves', abstract mathematical quantities with all the characteristic properties of waves which are related to the probabilities of finding the particles at particular points in space and at particular times. All the laws of atomic physics are expressed in terms of these probabilities. We can never predict an atomic event with certainty; we can only say how likely it is to happen.[8]

So, when we deal with quantum particles, we're dealing with probabilities. We can make educated guesses about where they might be or what they might be doing, but we can't know for sure until we actually measure them. The caveat, of course, is that when we measure them, the act of measurement affects the quantum system and effectively collapses the particle out of superposition and shows us an exact location. Given the nature of the observer's effect on quantum systems, we cannot measure a quantum system without effectively disrupting it, and thereby circumvent our ability to see it in its original state.

In order to work with this quirky situation where particles would only show "tendencies to exist" (probabilities of showing up in some region of space), our good friend Erwin Schrödinger came up with a novel mathematical tool called a "wave function." This wave function gives us information about the probabilities of finding a particle in different positions and states. The concept of "probability clouds" comes from this wave function. A probability cloud is like a foggy map that shows us where a particle might be located, with areas of higher and lower likelihood.

The uncertainty that arose from our inquiry into nature at the microscopic level cast a shadow on our attempt at constructing a world picture that was definite and certain. We learned that nature had some limitations built into it that would fundamentally restrict us from being able to speak with certainty about all aspects of what we observed. Moreover, we discovered that our own role as observers—despite our best efforts at objectivity—was somehow affecting our experiments in such a way that it was impossible to study nature without at the same time affecting it. These revelations and more upended our image of the real world during this time, yet it was Einstein's theory of relativity that changed our perception of space and time itself.

THE RELATIVITY OF SPACE AND TIME

The second great change to our interpretation of reality in this period came with regard to the nature of space and time. It was with Albert Einstein's theory of special relativity that we were able to determine that "space" and "time" were not actually absolute and unchangeable phenomena, but relative and malleable. Furthermore, they were not separated from each other, but instead existed in an interrelationship called a "continuum," one fabric to which both belonged. In order to make an accurate scientific measurement within space and time, one must account for both, translating from one "position" to another.

Up until this point, we believed that space was filled with a substance called the "ether" (sometimes called luminiferous ether), which basically filled space up completely, and through which light and other substances would travel. It was with Einstein's special theory of relativity that we (mostly) did away with this concept and began to consider space to be a "vacuum," or, in other words, filled with emptiness (it was not for a long time until we realized that even this wasn't exactly right!). Similarly, we believed time to be a constant and unchangeable phenomenon, the same for all observers anywhere in the universe. This was one of the most important changes to our perception of reality, as it was essentially the realization that time could be altered by speed, that helped us realize that our conceptions of space-time were not complete.

It was with Einstein's special relativity that we learned that light is both the fastest signal in the universe (and by signal we simply mean information traveling through space), and that its speed is constant and unchangeable. It was also around this time that we were able to determine the speed of light with a higher degree of accuracy—about 300,000 kilometers per second (186,000 miles per second)—denoted in scientific literature by the constant "c." One of the novel realizations of Einstein was that if light, the fastest signal in the universe, is traveling through space and is then perceived by another observer, let's say 1,200,000 kilometers away, they would not actually experience the phenomenon at the same time as one who is standing right next to it. It would take the light four seconds to travel from one observer to the other. In this way, the "time" at which this event occurred would not be the same for both observers. For the second party, the "event" is actually happening four seconds later.

Another famous example has to do with two observers, one standing inside of a train moving at a constant speed, and another standing outside of the train next to the tracks. Now imagine that in the same instant that the observer in the moving train passes the one who is standing next to the tracks, two bolts of lightning simultaneously strike the ground, one 20 km in front of the train, and the other 20 km behind it. For the observer on the ground, both lightning strikes happen simultaneously, but for the observer on the train, they will see the strike in front before the one in the back. Why?

For the observer on the ground, the strikes are the same distance from them, and the light travels at the same speed, so they reach them at the same time. For the observer on the train, the light from the lightning strikes must travel from where they struck the ground to him on the train, but because the train is moving in the direction of the one in the front, its light will reach him first. The light from the back will take longer to reach him, because he is in motion. What this basically means is that the "simultaneity" of events is not the same for everyone, but is instead relative to your frame of reference, from what position you observe. To say this another way, simultaneous events for one person who is still, will not be simultaneous events for someone who is in motion. This means "time" is not actually absolute but must take into account your frame of reference, both your position in space and your time at that position.

The idea of a "frame of reference" was hugely consequential to the science of this period. A reference frame is essentially a "starting point" for the purposes of any measurement. We must determine at which place and at what time a measurement is being made to know this frame of reference. It was with Einstein's special theory of relativity that time and space, which were considered separate phenomena up until this point, were joined into one. In his model, every "event" in the universe could be determined by four coordinates within this newly conjoined "space-time," three spatial coordinates (along the three spatial axes in three dimensions: up/down, right/left, and front/back) and one temporal coordinate (time). These four coordinates would correspond with a "point" in space-time. In order to make any accurate calculation of motion, one must take into account the frame of reference of each observer, and the speed and direction of their motion, to accurately "translate" between these two points in space-time (using a special method called the Lorentz transformation).

With Einstein's discoveries, we realized that the frame of reference of an observer is an indispensable piece of the puzzle in any scientific measurement. Time and space were not actually static and absolute, the same for all parties in all places, nor were they separate. Instead, they belong to the same fabric and are different for all parties, depending on their frame of reference and the speed and direction of their motion. These conclusions ultimately led to the other part of Einstein's theory, the *general* (versus special) theory of relativity, in which he demonstrated that the phenomenon of gravity was actually *bending* space-time, if an object was sufficiently massive.

The issue was that the special theory of relativity, while working perfectly for objects moving in a straight line and at constant motion, did not work when gravity was present or if the object was accelerating. Matter creates a curvature in space-time. This was proven via the orbit of Mercury. All planets orbited the sun in an ellipse. Mercury did also, but it also had a strange behavior called a precession that was a mystery to scientists at the time. Basically, not only was Mercury following its orbit around the sun, but it seemed the elliptical orbit itself was *also* rotating. By applying Einstein's new mathematics to the curvature of space and time via the effect of gravity on space, he was able, for the first time, to explain this strange behavior.

One of Einstein's other great contributions was in establishing that the speed of light— approximately 300,000 km per second—is constant. It does not change depending on the observer's frame of reference. Essentially, while light will bend in the presence of a gravitational field, it will not change its speed. In other words, light has the same speed in an accelerating frame of reference as it does in one that is at rest. This means that the speed of light in the presence of gravity will be the same as its speed in empty space.

Since the space traveled in a gravitational field is longer (given that the space is curved), in order for the speed of light to remain constant, it is *time itself* that must move slower in the gravitational field, relative to time in empty space. In other words, time speeds up proportionally with the curvature of space near the gravitational field compared with empty space, to keep the speed of light constant in both frames of reference. This is why time is considered distorted by gravity.

The idea that time itself could be distorted was novel! Up until this point in the scientific community, we mostly believed that time and space were constant and unchangeable phenomena, constant matrices within which the rest of "reality" existed. With Einstein's theory of special and general relativity, we put an end to the notions of absolute space and time. These were no longer absolute phenomena, but *relative*, with this relativity finding its home in the observer's "frame of reference."

THE OBSERVER

As it is with any revolutionary changes to our way of viewing the world, the experimental conclusions of the early 20th century were hard to digest for many who preferred to stick with a more traditional perspective. The Copenhagen interpretation of quantum mechanics—which explicitly acknowledges the role of measurement in physical systems—was especially challenging, given that it posed a direct challenge to any idea of the world based in strong objectivity. Heisenberg comments:

> … all the opponents of the Copenhagen interpretation do agree on one point. It would, in their view, be desirable to return to the reality concept of classical physics or, to use a more general philosophic term, to the ontology of materialism. They would prefer to come back to the idea of an objective real world whose smallest parts exist objectively in the same sense as stones or trees exist, independently of whether or not we observe them.[9]

What we discovered during this period in the early 20th century was that many of the notions of classical physics begun with Newton's Principles of Natural Philosophy and carried on by the bright minds who came after, held up as long as we were analyzing phenomena that were relatively close in size and scale to what we could witness with our physical eyes. But when it came to phenomena at scales significantly higher or lower in size, the rules of classical physics began to break down. Rules of causality and locality were called into question by phenomena such as entanglement. Strong objectivity and epiphenomenalism were called into question by superposition and the observer effect.

The issue of an "observer" became paramount. For quantum mechanics, this "observer" was seen in the fact that the smallest quantum objects like photons and electrons would *physically change their shape and behavior* depending on how we looked at them. In relativity theory, the observer became paramount because accurate measurements of space and time themselves were dependent on the frame of reference in which we made them, essentially, the time, place, and motion of the observer. In both cases, "the scientist"—who is supposed to be excluded from the phenomena they study—was now inextricably involved in interactions at both the biggest and smallest scales.

Not everyone accepted the implications of these results, including Einstein himself, who was famously opposed to the Copenhagen interpretation. First, because he was not comfortable with the inherent uncertainty in quantum systems (hence his phrase, "God doesn't play dice"). And secondly, because he believed that it was ultimately possible to get away from the viewpoint of the individual observer and fit it into a model that was universal and "public." American physicist Percy Williams Bridgman comments:

> Perhaps the most sweeping characterization of Einstein's attitude of mind with regard to the general theory is that he believes it possible to get away from the special point of view of the individual observer and sublimate it into something universal, "public," and "real." I on the other hand would take the position that a detailed analysis of everything that we do in physics discloses the universal impossibility of getting away from the individual starting-point. It is a matter of simple observation that the private comes before the public. For each one of us the very meaning of "public" is to be found in certain aspects of his "private." Not only is the starting-point in any scientific activity always private, but after it has emerged into the domain of the public, the story is not completed until we can return to the private from which we came. For the final test of scientific description or theory is that it enables us to reconstruct or to anticipate the immediate (private) situation.[10]

What Bridgman is saying here is that the "private" (our own individual and subjective perception), and the "public" (the "external" domain that we share with each other) have a relationship. What is public and external begins in our private experience, and when we make a scientific analysis, we must observe again from this private realm of experience to confirm its validity.

In fact, this was in line with Einstein's own observations. He has commented on the fact that science always begins and ends with the senses. That the conclusions of our intellectual analyses which arise from data we have gathered through the senses, must then eventually be confirmed via the senses at the end of the scientific process. The relationship between the sensory and the mental is central to the scientific process. In his own words:

> ... I shall not hesitate to state here in a few sentences my epistemological credo... I see on the one side the totality of sense-experiences, and, on the other, the totality of the concepts and propositions

which are laid down in books. The relations between the concepts and propositions among themselves and each other are of a logical nature, and the business of logical thinking is strictly limited to the achievement of the connection between concepts and propositions among each other according to firmly laid down rules, which are the concern of logic. The concepts and propositions get "meaning," viz., "content," only through their connection with sense-experiences.[11]

In this passage Einstein lays out his epistemological credo, which is simply his view on what constitutes valid "knowledge" or truth. He says that on the one side, we have the totality of sense-experiences, the experiences we have as individual observers of nature, and on the other hand, we have "concepts and propositions which are laid down in books," what could be called intellectual knowledge. While the relationships between the intellectual concepts themselves are the concern of logic, they do not actually take on any *meaning* until they are connected back to sense-experiences. In the beginning and in the end, sense-experiences are crucial. The aim of "conceptual systems," he says, is to coordinate as nearly as possible with the totality of our sensory experiences. He continues:

The degree of certainty with which this connection, viz., intuitive combination, can be undertaken, and nothing else, differentiates empty phantasy from scientific "truth." ... Although the conceptual systems are logically entirely arbitrary, they are bound by the aim to permit the most nearly possible certain (intuitive) and complete co-ordination with the totality of sense-experiences...[11]

The implication is clear. Science as it is cannot be done without the senses, even if we try to disembody the knowledge they provide and treat it as if it exists independently of the person who picked it up. After we have analyzed this data, we must return again to the senses to confirm whether our experimental hypotheses are true or not. Therefore, despite the fact that we do our best to exclude the observer—the subject—from our science and maintain a certain objectivity, the truth is, *it was never really absent*. The subject was always involved in science because science requires it at both ends of its process. It is only absent from the *model* it produces, not from its methods.

Science has taken us far. From our understanding of the atom and light to quantum physics and relativity, our knowledge has given us the tools to act on nature in highly sophisticated ways and create machinery that has transformed the landscape of our world. But the reality is, for all our knowledge, our most pressing issues remain. The absence of the observer in our scientific models has ensured a view of nature in which the reality of our subjective experiences has no place in our final assessment of things. This has guaranteed that we as observers of nature—as active participants in it—do not have any clear idea of our role in the cosmos, nor any clear roadmap of how to arrive at a peaceful and harmonious existence. To resolve this, we have to return to the roots of our ontology—our interpretation of reality—and examine our basic assumptions. The core of the issue comes down to this. If subjectivity is an objective feature of reality—*if it is always present*—are we truly being objective by ignoring it?

CHAPTER 8

MIND AND MATTER

As long as humankind has evaluated its position in the world, a distinction has been made between the realms of mind and matter. While in the modern world we take it as a given that the material world can be dissected and understood with respect to its mechanical components and the laws that govern its behavior, the universe of the mind is not as well understood. Even physicists like Max Planck, whose ideas shaped the very foundations of quantum mechanics, had strong feelings where it came to the mind's place within the greater architecture of the cosmos:

> As a man who has devoted his whole life to the most clear-headed science, to the study of matter, I can tell you as a result of my research about atoms this much: There is no matter as such! All matter originates and exists only by virtue of a force which brings the particle of an atom to vibration and holds this most minute solar system of the atom together. We must assume behind this force is the existence of a conscious and intelligent Mind. This Mind is the matrix of all matter.[1]

It is intuitive that mind has something to do with the dance of matter that surrounds us—that it is involved somehow. But what it is and how exactly it fits in is not so clear. To deepen our understanding of the mind's place within the overall order, we must begin by identifying it.

THE MIND AND THE BRAIN

To have this discussion properly, we must first make a distinction between "mind," the non-physical space where mental activity occurs, and the brain, the physical space where mental activity occurs. Both are associated with "the mental," but with the mind we are referring to mental *experience*, whereas with the brain we are referring to mental *physics*. What is meant by "mental physics" is just the physical processes that correlate to mental experience. On a physical level, the primary objects that correspond to mental experience in the brain are called neurons and neurotransmitters.

A neuron is a physical cell. It is like any other cell in the body, but with a few special properties. Firstly, neurons are associated mainly with brain activity (or "mental" activity, if you like) and also appear at times in other parts of the nervous system. They have some special physical attributes that set them apart from other cells, such as "dendrites," which are kind of like its antennae allowing it to *receive* electrical signals from other neurons, as well as "axons," which are kind of like transmission cables that allow it to *send* signals to other neurons (as well as muscles and glands). These special physical properties allow the neuron to perform its main job, which is to receive, process, and transmit information, something general cells are not designed for.

The second component in this mental dance is neurotransmitters, which are complex molecules (to reiterate, molecules are just two or more atoms come together to form a new substance). Unlike neurons, they are not living cells of the body. You can think of neurotransmitters as "the messages" that are being transmitted between "the messengers" (the neurons). Neurons communicate electrically over short distances, but use neurotransmitters to communicate over longer distances chemically. Some neurotransmitters excite neurons, making them more likely to send out messages, while others inhibit them, making them less likely.

There are several different types of neurotransmitters and each has its own specific role. Dopamine for example, is often associated with pleasure and reward, while serotonin impacts mood and sleep. When neurotransmitters have delivered their message, they're often reabsorbed by the neuron that released them. This process—known as "reuptake"—is targeted by some medications like antidepressants that alter the amount of certain neurotransmitters in the brain. Many substances introduced to the body from the outside such as drugs, other medications, and even food can affect this system, such as caffeine which acts as a stimulant by blocking a certain neurotransmitter that makes you feel tired.

MIND, THE SCIENCES, AND REALITY

This complex system of neuronal activity comes together to form a network of interactions at the physical level that supposedly correspond to the suite of mental experiences that occur within the subjective observer's mind. New technology using artificial intelligence (AI) is attempting to map neural activity in the brain and correlate it with certain thoughts. The field of prosthetics for example is working on technology that will use AI to map the parts of the brain that light up when a subject is thinking about moving a limb like their arm. Given AI's strength with pattern recognition through machine learning, it can quickly learn and adapt through repeated testing and begin to "predict" with greater and greater accuracy when someone is having a "thought" related to this type of activity.

As remarkable and exciting as this technology is, it is still not intended to make any sort of ontological statement about reality. We are not saying anything about what thought is or where it comes from. For the purposes of the development of this technology, it doesn't really matter. What matters is that when the subject focuses on moving their arm, the machinery registers the way their brain changes in response to that intention. Over time, we can develop statistical models that give us a general idea that when a person's brain changes in such and such way, they are trying to move their arm. The machine can then respond to this "thought" and react accordingly. These sciences are beautifully practical. Unlike physics or metaphysics though, they are not meant to make ontological statements, to discern universal or natural laws, the way reality works or the rules that govern it.

Even the sciences that are explicitly oriented towards studying the brain and the mind, fields like psychology, psychiatry, and neuroscience for example, are not designed to examine questions about the role and nature of the mind in the cosmos—within the fabric of reality itself. These sciences are more practical in nature, oriented toward examining the manifestation of the mind in the physical world within our bodies and our daily lives. Psychology for example, examines the realm of thought in a more anecdotal way. It relies in part on first-hand experiences of its subjects to identify patterns in thought and behavior that can lead to a deeper understanding of the nature of our experiences.

Neuroscience goes a layer deeper, studying the biological machinery that makes psychological phenomena possible, including the aforementioned neurons and neurotransmitters. Using technology like MRI scans and electrophysiology, neuroscience aims to map out how different parts of the brain correlate to our mental lives. Psychiatry—which has the unique authority to prescribe medications—blends the insights of psychology with a medical approach focused on diagnosing and treating mental disorders, thus merging chemical interventions with psychological understanding. All of these fields, while tremendously complex and valuable in their own ways, are not focused on understanding the role of mind, thought, or mental experience within the fabric of nature itself.

Physics—as well as other fields such as chemistry, astronomy, and mathematics—are sometimes referred to as "exact" sciences. While this terminology is falling out of favor, given that no science can really be said to be exact in the absolute, they are pointing to the fact that they are capable of reproducing the physical situation with a high degree of precision. Loosely speaking, the exact sciences (also sometimes called the "hard sciences") don't require us to understand human psychology. This is because psychology doesn't allow us to observe entities or patterns in the same way we can with physical ones.

At the core of the problem is that "mind"—as opposed to the brain—is not a physically manifest entity with properties that are discernible with our five physical senses. Matter on the other hand is physical by nature, and therefore can be measured in the usual way. Given that the realm of mind does not possess these same types of discrete, visible qualities, it cannot be measured with quantitative instruments. Whatever is not quantitative in this way cannot be dealt with mathematically. For this reason, mind and matter—as understood through the concept of energy—cannot easily coexist within the current scientific framework.

Schrödinger in his wonderful little book *What is Life? With Mind and Matter* refers to this problem. He comments on the price we pay for removing ourselves from the world picture in order to do science in our chosen way. This includes the essentially "dead" form it takes, devoid of any of the sensual qualities that make up the world of our experience, as well as the absence of any adequate way to represent the interaction between mind and matter:

> ...For the moment let me just mention the two most blatant antinomies due to our awareness of the fact that a moderately satisfying picture of the world has only been reached at the high price

of taking ourselves out of the picture, stepping back into the role of a non-concerned observer. The first of these antinomies is that astonishment at finding our world picture 'colourless, cold, mute'. Colour and sound, hot and cold are our immediate sensations; small wonder that they are lacking in a world model from which we have removed our own mental person.

The second is our fruitless quest for the place where mind acts on matter or vice-versa, so well known from Sir Charles Sherrington's honest search, magnificently expounded in *Man on his Nature*. The material world has only been constructed at the price of taking the self, that is, mind, out of it, removing it; mind is not part of it; obviously, therefore, it can neither act on it nor be acted on by any of its parts.[2]

British neurophysiologist Sir Charles Sherrington (whom Schrödinger mentions in this passage) dealt extensively with this issue. Being a physiologist, his focus was not just on physics, but on the body, and more broadly, on life itself. One cannot work with a living organism without acknowledging to some degree the interplay between it and its material environment. Being active in the early to mid-1900s, he was witness to the way in which physics had continued to reduce the basic physical elements of our world—of life itself—down to finer and finer parts. But in the interaction between these parts, there was a gaping hole. The very mind that did the analysis of this world was absent from the picture it produced.

Life …as knowledge got nearer to it, had resolved itself into a complex of material factors; all of it, except one factor. There science stopped and stared as at an unexpected residue which remained after its solvents had dissolved the rest. Knowledge looking at its world had painfully and not without some disillusions arrived at two concepts; the one, that of energy, which was adequate to deal with all which was known to knowledge, except mind. But between energy and mind science found no 'how' of give and take. It could meet all efficient causes but not a final cause. To man's understanding the world remained obstinately double.[3]

Close to the core of this issue is the concept of energy. Supposedly, everything in the universe is composed of energy. This even includes matter itself, brilliantly expounded by Einstein's famous equation $E=MC^2$, which simply states that energy is equal to mass (sometimes thought of as the amount of matter in an object) times the speed of light squared. If we take this concept—that everything including matter is energy—and extend it further by considering the law of conservation of energy, which states that energy can neither be created nor destroyed, only changed in its form, we get a picture of the world where there is a fixed, finite amount of energy that is in an endless dance of transformation from one form to another. So in this dance, where is mind?

Among all the different varieties of energy we have identified in our models (thermal, radiant, chemical, nuclear, etc.), there is no such thing as "mental" energy. Yet if mind is interacting with matter in any way at all, then it must exchange energy. How is it then that mind interacts with matter? Coming back to the fact that the measurement of an *experience* is not the same thing as the measurement of an *object*, we see that there is a fundamental inability to include mind in the energy scheme of our modern models. Sherrington continues:

> Conscious experience seems refractory to measurement in terms of itself. We cannot say that in experience one light has twice the brightness of another. The terms in which we measure experience of a sound are not terms of experience. They are terms of the stimulus, the physical sound, or of the nervous or other bodily action concomitant with the experience. Or the measurement may be in terms of some motor act, which is taken as an outcome of the mental experience. Mind, if it were energy, would be measurable quantitatively. For quantitative measurement of the mental we resort to the energy-scheme. But the validity of that resort is questionable. The search in that scheme for a scale of equivalence between energy and mental experience arrives at none. The two seem incommensurable.[4]

Despite the fact that any attempt at finding equivalency between mind and energy (and by extension, mind and matter) have come to nothing, our materialist philosophy still tends to equate the two. This leads to a question often at the back of people's minds when it comes to this arena: Is mind just matter?

IS MIND JUST MATTER?

There are many ideas about what mind is and how it fits into the overall order of the cosmos. René Descartes' idea that mind and matter are two fundamentally different spaces, named "Cartesian dualism" after him, was an idea born at a time when material scientists were still often active metaphysicians. After the Scientific Revolution of the 16th century, it became less and less common for scientists, those involved in the study of the material world through quantitative methods, to be involved in trying to solve ontological questions. This type of work dealing with the nature of reality and our experiences was slowly relegated to the realm of philosophy. What we don't often think about are the consequences of this oft-unspoken development.

Quantitative measurement is well-suited to the study of material objects, but doesn't do so well when it comes to phenomena internal to our experience. Given the material bent of our science and the way it has influenced our collective thinking about nature, we tend to assume that "mind" or "mental experience" is simply produced by the brain. Our inclination is to assign a material cause to this phenomenon, as we assign material causes to everything else in the world picture our science produces.

A good example of this is the "identity theory" which simply states that all mental processes are identical to physical processes in the brain and nervous system. Since neurons are the "cells" of the brain and facilitate the interactions that make brain activity possible, they become the root cause of what we call "conscious awareness." French theoretical physicist Bernard d'Espagnat lays this out:

> The view here called, for brevity's sake, the "identity theory" consists in asserting that, in the last analysis, any genuine sensation— that is, any "becoming aware" and, by extension, perhaps even any thought—is finally identifiable to some material structure internal to or involving neurons.[5]

According to this theory, for every mental state there is a corresponding physical state in the brain, and these states are one and the same. The implication is that the two are not merely correlated, but *ontologically identical*. They are the same thing looked at in two different ways: one from the inside (subjective experience) and one from the outside (objective observation). The question of which is acting on which, mind on brain or brain on mind, becomes irrelevant within this framework, because for all intents and purposes, they are identical. In other words, mental phenomena aren't *caused* by brain states, they *are* brain states.

This takes us to the most common assumption within our collective thinking about what the mind is. *It's just matter.* Subjective experience is a consequence of the material constituents that make it up (mainly neurons). The implications of this way of thinking on our logical and ontological framework are important and far-reaching. If mind is simply matter, or better yet, if mind—the internal experience of the conscious observer—is *caused* by matter, then there are a lot of other things that must logically follow from that. Exploring this line of reasoning is important, because if our assumptions are based on faulty logic, it places our entire ontological framework on shaky ground.

THE MIND-MATTER CONNECTION
AND THE CAUSALITY OF CHOICE

Part of the problem with this purely materialist framework has to do with the fact, highlighted above, that we can look at things in two different ways: "from the inside" and "from the outside." While the identity theory wants to say that mind and matter are identical, it does so while in the same breath saying there is a division, even if only in the way we view things, between subjective experience (inside) and objective observation (outside). This implies (even if it is not stated directly) that there is an ontological difference between the two. At least from our own point of view—that of the observer—this is very obvious. We have internal processes dealing with our own personal thoughts, feelings, and experiences, and these stand alongside an "external" reality that we share with others, and seems to exist in a more "objective" way, independent of our personal experiences.

The reason that the "inner" or "subjective experience" has to be mentioned at all, even in a materialistic theory like the identity theory, is because it is verifiable from our experience. It is a central component of the reality we witness and therefore must be accounted for in some way. Since science's way is to objectify ("to make an object out of") everything, when speaking of the connection between mind and matter, it aims to reduce or eliminate these subjective elements as much as possible. The issue, of course, is that the mind is not an object. It cannot be measured or quantified in any physical way that renders it subject to our scientific instruments. What is more problematic though, and seldom stated explicitly, is that *the mind always has a subjective component.* This confirms itself in every instance and example we can observe in nature. There is never a time we observe mind without it belonging in some way to a subject—to an observer. Despite the fact that we prefer to view nature from a viewpoint in which we remove our own mental person, we must remember that this is only an idealization, not an observable truth.

Moreover, when we equate the subjective and the objective in the way typical of our materialistic theories, we brush over issues of causality. Is the inner (mind) affecting the outer (matter)? When we equate the two in our models, we get to conveniently sidestep the question. Causality refers to the order in which things happen. One thing can *cause* another and therefore comes earlier in the chain of events. For example, when we turn the key in the ignition of our car, the engine starts. The turning of the key causes the engine to start, and it's clear which comes first: we can't have a running engine without first turning the key. There is a reversible interaction in time. Beginning from the present moment and moving backwards, we can piece together a series of events to understand what events are causing other ones. Understanding root causes is at the heart of science—in physics certainly, but even more so in metaphysics.

When we "act," we seem to do so from an internal place (mind), on or through the body (matter). Sherrington says:

> My mind seems to act on my 'material me' when at breakfast I lift my coffee cup with intent to drink. I infer a like situation in the chimpanzee when he peels his banana before eating it. Reversible interaction between the 'I' and the body seems to me an inference validly drawn from evidence.[6]

This connection between mind and body is such that the content of our thoughts can affect the body even down to the cellular level. Being in a state of repeated mental stress or anxiety can cause oxidative damage to physical cells and DNA. When we exercise at the gym, mental focus on muscle groups can cause the body to "activate" and use those muscles in a more targeted way. Repeated cultivation of a relaxed or spacious mental state (through activities like meditation) has been shown to have a profound effect on brain activity, especially over time. The effect of mental states on physical cells is well-documented in fields like psychology, psychiatry, and neuroscience, but the ontological questions it poses for the connection between mind and matter in a broader cosmological sense are seldom considered.

At the heart of the issue is the causality of choice. When we choose to act, where is the source of this action? Evidence suggests that it comes from that non-physical space, the "internal" we mentioned earlier. The salient point is that—*choice finds its origin with the subject, not the object.* This poses problems for a strictly materialistic interpretation of physics, and more broadly, for any ontological model which aims to reduce or eliminate subjectivity. This kind of model becomes limited in its explanatory power when confronted with phenomena that have observable qualities and characteristics that are not quantifiable.

This is the core of our problem. There is an observable interaction between us as conscious observers and the object of our observation. There is an interaction between the "I" who does the perceiving and the object of perception. We take action in the world and make choices. These choices supposedly affect the material world. Were it not the case, we would not make them in the first place. Yet, within the ontological model created by our current methods, no such choice exists. There is no way for mind to affect matter. Clearly, there is something missing. To move forward, we must find the link that bridges these two seemingly disparate realms. We must identify what is missing from our current models that may be creating this disconnect. The more we can identify areas where our current efforts have overlooked, the greater the likelihood we will be able to reconcile any conflicting data and create a model that is whole.

CHAPTER 9

FEELING AND
THE SENSE OF TOUCH

efore we continue with our review of the science, let us recap some of what we've learned. Our thinking about the nature of reality has evolved over thousands of years, beginning with the idea that nature was governed by deities, subjective entities with control over the forces of nature. Starting with the Ionian Enlightenment in the sixth century BC, our thinking began to shift toward the idea that nature was actually governed by objective processes we could ourselves discern. This led to the field of natural philosophy which, over the course of the next 2,000 years, created a large body of knowledge from philosophers all over the world whose ideas influenced everything from faith to technology to politics.

This reached an apex somewhere around the sixteenth to seventeenth century when much of the recovered knowledge of the Greeks, which had returned to Europe through the Arab Empires that had preserved it, began to influence European thought anew. The faith-based religious environment of that period began to give way to the old reason-based philosophies of the ancient world, which transformed our worldview yet again. This came to a head with the Scientific Revolution and the Enlightenment, during which the empirical and rational schools of thought developed. This eventually culminated in a new "scientific method" em-

phasizing empirical observation, experiment, and quantitative methodologies (including the use of mathematics) to discern the core principles
of nature.

During this period, subjectivity and objectivity began to diverge from
each other in our collective inquiry into the nature of reality. Analysis
of physical, quantitative, and general aspects of reality were emphasized
over personal, subjective, or qualitative ones. This divergence eventually
gave rise to a schism in natural philosophy splitting it into two distinct
branches: philosophy and science. While both were reason-based, science dealt with the objective analysis of mostly physical systems using
the scientific method, while philosophy dealt with everything else. This
divergence held for many hundreds of years, beginning with the birth of
classical mechanics until the turn of the twentieth century when quantum
mechanics and relativity theory began to reintroduce the observer back
into our world picture. While these theories introduced ideas such as "the
observer effect" and "frames of reference" which implied the necessity
of a subject in any final assessment of things, they stopped short of mentioning subjectivity directly. Since then, the realms of mind and matter
have remained mostly separate from each other, without any clear way to
reconnect them.

THE INTUITIVE AND THE RATIONAL

There's one aspect of the science that emerged from our methods during
this time that we don't often consider. Of all of the information that is
available to us to study empirically, all of the data that reaches us through
the senses, our science is focused mostly on two: sight and sound. The
vast majority of the science we do is focused on data that can be gathered
through our eyes and ears, and which can then be analyzed with reason
(the mind). When it comes to our other senses, especially smell and taste,
but also to a lesser extent touch, this information is greatly deemphasized
(if not outright ignored) in our models.

When it comes to our experience of the natural world, consider that
some senses are better at discerning certain phenomena than others. Let's
take temperature for example. With the sense of sight, we can tell to some
degree whether something is hot or cold. Cold things tend to turn white
and get more solid. Sometimes there is the formation of ice. Hot things
tend to get brighter and glow (typically yellow-orange). There is some visual evidence of temperature, but with our eyes alone, we cannot truly be

sure. Hot things do not always glow bright and cold things do not always turn solid or white. The eyes can *help*. They can give us some indication, but they are not totally reliable. Neither smell nor taste can help us here either. The only sense that can tell us with certainty whether something is hot or cold is *touch*. We can *feel* something's temperature, regardless of what it looks like.

Let's take speed as another example. Imagine we are in a vehicle that begins to accelerate. Visually, we are able to see that we are in motion and relatively measure our speed by the rate at which the things around us pass by. To a degree, we can also *feel* speed. Especially at high velocities, we can feel g-force (gravitational force) on the body. The faster we move, the more obvious this effect is. Similarly, higher and higher speeds create a visual phenomenon of blurred vision on the sides or periphery of our movement and clearer vision of the objective or destination in which we are pointed. The senses of touch and sight are useful in gauging speed, but our senses of smell, taste, and hearing cannot really help us here.

These are only two of many examples. There are various phenomena in the world of our experience, and not every sense can perceive all of them. Moreover—and crucially—*not all of them are physical*. There is also the whole gamut of experiences we have in our *inner* lives that form a different spectrum of phenomena. We can see, hear, and feel a broad spectrum of things that may not be physically visible to others in our environment or measurable with quantitative instruments, but which form a part of our internal experience. For example the "visualization" of a bird in one's mind. The sound of our grandmother's voice from early childhood. The remembrance of a lover's touch. The smell of someone's scent on an old t-shirt. This inner aspect of reality extends far beyond simple physical experience to a suite of inner mental and emotional experiences that live far outside the reach of our current scientific methods. This is the case, even though they are critical to the experience of reality as we know it, from the perspective of the individual observer.

Human emotion—what we often call *feelings* in our language—is not something we can see, smell, taste, or hear, but it is definitely something we can *feel*. The experience of *feeling* the world is critical to our reality. It is often the very thing we associate with being human. In our science fiction movies it is often the wide gamut of human emotion and our capacity to feel that we use to distinguish us from the portrayal of extraterrestrial races. We know from within our experience that this capacity for feeling is critical to our experience of the world—even central to it—but we have been content to leave it out of our models, because it is fundamentally qualitative and unquantifiable. We are resigned to the idea that it doesn't have discernible cause-and-effect mechanisms, general principles, or any internal structure we can build a model from.

The emotional experience is primarily perceived through the sense of *touch*. We tend to think of touch as something that happens on the skin, but consider the fact that when your kidney hurts or your heart is aching, you are just as able to feel what is happening *inside* your body as you are what is happening outside of it. Feeling permeates the whole body. Out of the five senses, it is the only one which does not depend on a specific organ, and the only one that can potentially be present in every cell.

Finally and most importantly, giving evidence to the primacy of touch among the five senses, *it is the only sense we cannot totally lose*. It is possible to lose our sense of smell, taste, sight or our hearing, but we cannot, totally and completely, or as a permanent condition, lose our sense of touch. It is the way by which we feel. It is connected, in the most fundamental way, to sense itself.

Often times when we say that we "sense" something, we are saying that we *feel* it. There is contact with another reality. Touch implies contact. It is the point at which two or more bodies interact. To sense, to feel, and to touch are very close to each other. They are all describing a direct, visceral contact between entities. This is quite opposed to the indirect, impersonal, and "disembodied" nature of intellectual knowledge.

Consider what it is like picking out a sweater at the shopping mall. Sometimes you pick up the sweater and touch its material. There is information about the sweater you receive through touch. This is *sensory* knowledge. It is distinct from the type of knowledge that reaches you through your intellect, what could be called intellectual knowledge.

Intellectual knowledge is indirect and impersonal. Sensory knowledge is direct and personal. The sensory and the intellectual form a sort of duality in the nature of all the information that we receive about the world. You could call information that comes through sense and feeling "intuitive," while information that comes through the intellect "intellectual" or "rational." These line up nicely with the subjective (intuitive/sensory) and objective (rational/intellectual) dualities we have already discussed. In truth, both sides of this duality are necessary for an understanding of reality as a whole.

QUALIA AND QUANTA

The duality between the intuitive and rational is closely related to the duality between the qualitative and the quantitative. Quantitative data (sometimes called "quanta") is numeric, countable, and discrete. Because it is "quantized"—with clearly defined units—it can be dealt with mathematically. Qualitative data (sometimes called "qualia") is indiscrete, sensory, and intuitive. It has a nature that is better understood from the "inside-out," from the frame of reference of the observer. As opposed to the nature of quanta, *qualia requires an observer*.

When we look at the sum total of the qualitative data that is available to us to study, we focus quite extensively on sight and sound (and to a lesser extent touch). Our quantitative instruments have been built to extend our natural abilities in these areas. Telescopes and microscopes allow us to see much closer and much further away than our physical eyes would allow. We have built vibrational sensors to detect sound far below what our ears can detect. We have built pressure and temperature sensors to measure these phenomena in a way that would be difficult or even impossible with our sense of touch alone.

This instrumentation has allowed us to reproduce a picture of the physical world with much greater precision than would be otherwise possible. Still, it leaves out quite a lot that is central to our lived experience, especially as it relates to kinesthetic (touch-based), olfactory (smell-based), and gustatory (taste-based) data. Incidentally, it is these data that are closer to those experiences that are often considered part of our "lower" or animalian nature, what could be called the "instinctual," standing apart from the intellectual and the intuitive.

Setting the sense of touch aside—it being primary to sense itself—it is often the senses of smell and taste that come first in the evolutionary process. The motivation to eat, especially something that brings pleasure to the senses (taste and smell primarily), is one of the greatest motivating factors in nature. Moreover, it is often through the sense of smell that many creatures "see." More often even than the sense of sight, the sense of smell is used by many creatures to see and perceive the world around them. This is true even and especially in the insect kingdom. Smell and taste are much closer to being pure qualia than any of the other senses.

This is all to say that certain senses lend themselves better to a quantitative treatment of reality. Through sight and sound, and to a smaller degree through touch, it is easier to sense and perceive discrete, countable phenomena. Data must come in this way if we are going to subject it to mathematical treatment, and is easier to deal with from the intellect, which loves to see things in their parts. The senses of taste and smell however (and again, to a degree the sense of touch), are not so easy to "break into pieces." They are not as easy to *quantize*. Therefore, while all of the senses have a qualitative component that delivers direct, empirical knowledge to the observer, only some of them can deliver information that can be easily fit into our quantitative frameworks. Considering this, it is no wonder that our science has become so sight and sound heavy, as these are the senses that deliver qualia which are easiest to *convert into quanta*, and which can then be dealt with mathematically.

If we take a step back and look at the whole display of nature and all the experiences that we can draw data from, it is obvious we have not really been looking at everything. Our inquiry has been focused in a certain way. We cannot deny that our methods have been successful for the aspects of reality that we have chosen to investigate, but to look at that and call it everything would be a gross oversight. While we have found success via these methods, we must acknowledge what is missing from our models, so that we can expand our inquiry to properly encapsulate the whole. As Einstein once said:

> It can scarcely be denied that the supreme goal of all theory is to make the irreducible basic elements as simple and as few as possible without having to surrender the adequate representation of a single datum of experience.[1]

When we look at our inquiry into reality from a bird's eye view and within the context of our discussion, it is easier to see which datum of experience has not been adequately represented. If our goal is to understand *reality*, we must reevaluate the nature and scope of our probe into nature. We must evaluate the data that must be accounted for in order to represent the reality of our experience in its totality. As long as our inquiry is limited only to certain methods, to certain senses, or to certain experiences, we cannot confidently proclaim that we are speaking to the whole.

THE MEANING AND PURPOSE OF THE WHOLE DISPLAY

Since the Industrial Revolution, our science has had great utility. Our understanding of physics allowed us to build machines that served to popularize science in the eyes of the masses. The institution of science, which was so much interested in truth and the nature of reality for the greater part of two millennia, started to be concerned more with what nature could *do*, rather than with what it *was*. Add to this the exclusion of the humanistic factors that make up our own lives, and we ended up with a cold, mechanistic science that looked more and more utilitarian—like a tool—rather than something possessed of any kind of meaning.

When we examine any entity, whether organic or mechanical, there are always two components to it. There is its internal composition, all of the parts that make it up, then there is the thing itself, the entity as a whole. When we examine the parts, such as the central processor of a computer for example, we see a particular function. Processors contain millions of transistors that switch on and off to represent binary data (1s and 0s). This is its function. But what is its *purpose*? *Why* does one need to register binary data? What greater role does it serve? When we examine the processor within the context of the whole computer, its role and purpose within the greater orchestration of the parts is much clearer.

Parts give us an idea of composition and function. They could be called "subtotalities," things that are complete in themselves, but which are part of a greater structure. For example, an engine is whole in and of itself, but its purpose—its reason for being—is revealed when seen in context of the vehicle it powers. A lugnut has its own form and function, but it takes on meaning when we see how it fits within the mechanical structure it is part of. In order to understand the purpose of a thing, we must understand it in the context of the greater whole to which it belongs. Otherwise we get stuck at the level of parts, and our focus becomes functional and utilitarian, but ultimately, the greater purpose—*the meaning*—is lost. It is purpose that speaks more to the *why* of something, function speaks more to the *what or how*. While function reveals itself in the composition of the parts, *meaning and purpose are revealed in wholeness*.

What is important to note is that it is the senses that tend to see things this way—in their wholeness. Think about the visual of a sunset, the sound of music, the taste of french fries, or the smell of a rose. It is natural to the senses to take things as they are, to take them in their entirety. But as soon as we begin to analyze the sunset, to dissect the sound of the music—to pick them apart and to try to understand their composition—we engage the intellect. We enter the intellectual space of the mind. Given the emphasis on the intellect in our scientific model, this distinction has an important bearing on the way we do science, and the picture of reality that emerges from it.

Our science is much more interested in the parts—*in taking apart the world*—than what it is in its wholeness. To a degree, this is what we have done with the whole cosmos. We have come to study and understand the universe in specialized pieces, and in that we have been very successful.

But when it comes to the meaning of the whole display, we have remained resolutely in the dark. From our current outlook, it doesn't seem to have any meaning at all. It seems to be more like a machine, just running on its own, indifferent to whether there is such a thing as a "subjective observer" or not. But I ask you, does it seem logical that nature would facilitate something as intricate and nuanced as human life without reason? Does it sound logical that something as rich and complex as human experience would be brought about over and over again throughout history in an infinite variety of ways only to disappear into nothingness without any discernible purpose?

We elect to set aside elements of our observable reality that pertain to our subjectivity, then are very surprised that on the other side of our analysis, the subject is absent. When it seems to find its way back in, we use highly unscientific terms like "spooky" or "mysterious" to refer to it, content to resign ourselves to the idea that "there are just some mysteries we will never figure out." The irony is that we used to speak this same way about lightning in the sky and the motion of the sun and moon. To emphasize the point, we return to the thoughts of luminaries Schrodinger and Sherrington:

> . . . I believe it to be true that I actually do cut out my mind when I construct the real world around me. And I am not aware of this cutting out. And then I am very astonished that the scientific picture of the real world around me is very deficient. It gives a lot of factual information, puts all our experience in a magnificently consistent order, but it is ghastly silent about all and sundry that is really near to our heart, that really matters to us. It cannot tell us a word about red and blue, bitter and sweet, physical pain and physical delight; it knows nothing of beautiful and ugly, good or bad, God and eternity. Science sometimes pretends to answer questions in these domains, but the answers are very often so silly that we are not inclined to take them seriously.[2]

> Natural Science thus attempts to withdraw itself from much that is human. ...Observing the perceptible, the scientific observer tries to divest himself of 'causes', 'forces', 'absolute time', 'absolute space', 'beginnings from blank', 'endings in nothing', 'ultimate reality', 'life', 'death', 'personal Deity', to say nothing of 'good', 'bad', and 'right', and 'wrong'. ...After all it suffers from the old complaint which Socrates preferred against Anaxagoras in the Phaedo. Man as scientific observer becomes an instrument for pointer-readings in the hand of

a disembodied intellect.[3]

It is imprudent and shortsighted to call something "mysterious" simply because it doesn't fit with preestablished ideas or frameworks. This attitude is contrary to the very spirit of science itself, which aims to understand, as accurately as possible, the nature of the world, and our own place within it. While the former, the nature of the world, has grown exponentially in the last half millenia, the latter, our own place within it, has been content to be forgotten at the edges of our collective knowledge.

If we have not yet been able to resolve the nature of mind or many of life's other biggest questions with our existing models or modes of thinking, it might be time to consider that maybe it is our methods and models that need to evolve. Rather than rejecting or setting aside those elements that are human, it may be wiser to consider that we may not yet be looking at things in the right way. Perhaps our whole reality has to be considered *together*, rather than in its parts.

Finally, we must admit that a cosmological model that requires reality to be considered from the perspective of a disembodied intellect (an intelligence that exists irrespective of any individual perspective) may be problematic if in fact that "body" (entity/subject/consciousness) is itself integral to the whole display.

There is one area of science where emphasis on the whole is more obvious, and that is in biology. In the world of the living organism, it is more evident how each part works together for the sake of the whole. Moreover, transition to the world of the living helps us to better understand the nature of reality, not only within the context of its physical components, but also within the context of life. An examination of life alongside our discoveries in physics can help us to have a more complete conversation about, not only the nature of the cosmos, but also our role within it.

PART III

THE ORGANISM

&

THE MACHINE

CHAPTER 10

THE IDEAL AND THE ACTUAL

The period following the great discoveries of the early 20th century was marked by a transition from a focus on natural science to technical science. While natural science is concerned with understanding the nature of the world as it is, technical science is more concerned with the application of scientific principles in the production of tools and technologies. This shift was aptly captured by Heisenberg:

> ... Natural science proceeded to get a clearer and wider picture of the material world. In physics this picture was to be described by means of those concepts which we nowadays call the concepts of classical physics. The world consisted of things in space and time, the things consist of matter, and matter can produce and can be acted upon by forces. The events follow from the interplay between matter and forces; every event is the result and the cause of other events. At the same time the human attitude toward nature changed from a contemplative one to a pragmatic one. One was not so much interested in nature as it is; one rather asked what one could do with it. Therefore, natural science turned into technical science; every advancement of knowledge was connected with the question as to what practical use could be derived from it.[1]

The science that emerged after relativity theory and quantum mechanics was characterized more by refinements to our current science than any radical redefinitions to our worldview or general cosmological outlook. This was an important point in the history and evolution of our study of nature. When we turned to technical and industrial applications of science, the philosophical backbone that led to scientific advancement was made subservient to those things that could have practical application in the real world.

Consequently, most of the changes to our science in recent centuries have had more to do with the practical application of scientific principles and the invention of new technology than any total revolutions to our way of thinking. Technology has allowed us to profoundly change the way we live, but the way we see life and the reality of the cosmos has more or less been colored by the same concepts.

So far we have discussed the duality between the senses and the intellect, what could also be described as the duality between feeling and thinking. The feeling aspects that are not physically tangible nor measurable with physical instruments are refractory to study with quantitative tools. There is no way to fit the world of felt content and subjective experience in our current cosmology. But the phenomenon of life is not visible only in personal experience. It is also visible *physically* in the world of biology and the living organism.

While it may seem intuitive that the world of biology must also obey the laws of physics, that biology is subservient to physics, there are actually many aspects of the biological world that do not fit with our current physical models either. When we consider the organic elements of the physical world and place them alongside those that are purely mechanistic, between those which are *biological* alongside those which are purely *physical*, another duality begins to form. This is the duality between *the organism* and *the machine*. Between that which is *alive* and that which is *dead*. Between that which is *animate* and that which is *inanimate*.

LAPLACE'S DEMON

In a way, the duality between a living thing and a dead thing is the same as the duality between the "subject" and the "object." The subject *experiences, feels*, or can otherwise *sense* on some level. It has its own capacity to *choose* based on its biological imperatives, its "self," if you want. An object on the other hand is just a "thing," that which has no life and does not feel or experience. It is "dead" and simply obeys mechanical laws.

Now what happens when the mechanical conceptions of nature get the upper hand? What happens when organic, subjective, or experiential aspects are set aside, and physical laws are considered the primary ones needed to explain everything? When we imagine the world as a perfect machine with perfect order and no random or unpredictable factors? We get Laplace's demon.

Pierre-Simon Laplace was a French scholar and polymath from the late 18th and early 19th century who did important work in the fields of engineering, mathematics, statistics, physics, and astronomy. What is known as Laplace's demon came from a philosophical essay originally published in 1814 in which he articulated the natural consequence of this way of thinking about the world. In his own words, it goes loosely as follows:

> We ought then to regard the present state of the universe as the effect of its anterior state and as the cause of the one which is to follow. Given for one instant an intelligence which could comprehend all the forces by which nature is animated and the respective situation of the beings who compose it—an intelligence sufficiently vast to submit these data to analysis—it would embrace in the same formula the movements of the greatest bodies of the universe and those of the lightest atom; for it, nothing would be uncertain and the future, as the past, would be present to its eyes.[2]

This is sometimes known as causal determinism (or scientific determinism), as we have mentioned before. To clarify, this viewpoint goes more or less like this. Since the universe is governed by strict laws that we can figure out, an intelligence large enough to contain all the laws and the positions of all the bodies in the universe at the same time would be able to perfectly predict what would happen from the present moment all the way until the end of time, and similarly, would be able to reverse engineer the current configuration all the way back to the Big Bang. In other words, since all events in the universe obey perfect laws of cause and effect, everything has already been determined.

What is interesting about this idea is that it makes no distinction between physical objects and biological life. It considers the behavior of inanimate objects and the behavior of man and animals as one and the same. For commentary on this we turn to Danish theoretical physicist Neils Bohr, a contemporary of Einstein and one of the most famous physicists of this period:

In order to present the situation in physics as clearly as possible, I shall start by reminding you of the extreme attitude which, under the impact of the great success of classical mechanics, was expressed in Laplace's well-known conception of the world machine. All interactions between the constituents of this machine were governed by the laws of mechanics, and therefore an intelligence knowing the relative positions and velocities of these parts at a given moment could predict all of the subsequent events in the world, including the behaviour of animals and man.[3]

The implication here for living things is that *there is no such thing as choice*. If everything I do follows precisely from the laws of physics, then the idea that "I," as a subjective observer, can affect my environment (or even my body) is really just an illusion. Precisely what I would do, including all the thoughts I would ever have and all the actions I would ever take, were already determined at the outset one instant after the Big Bang. All living things are essentially cogs in a giant world machine that ticks objectively according to strict physical laws irrespective of the subject, their experience, or their supposed "choices."

Along this line of reasoning, what happens in the universe at any given moment follows exactly from the previous moment, and laws of cause and effect determine everything. Therefore, the idea of "conscious choice" is really an illusion. We may be experiencing the world, but we are not affecting it. This is the consequence of a way of thinking in which we consider physical laws to be the only ones that exist. Erwin Schrödinger, lucid about so much when it came to these relationships between life and physics, also made comment on this:

The world-model consisting of atoms and empty space implements the basic postulate of Nature being understandable, provided that at any moment the subsequent motion of the atoms is uniquely determined by their present configuration and state of motion. Then the situation reached at any moment engenders of necessity the following one, and this the next following one, and so on for ever. The whole going-on is strictly determined at the outset, and so we cannot see how it should embrace also the behaviour of living beings including ourselves, who are aware of being able to choose to a large extent the motions of our body by free decision of our mind. If then this mind or soul is itself composed of atoms moving in the same necessitous way, there seems to be no room for ethics or moral behaviour. We are compelled by the laws of physics to do at every moment just exactly the thing we do; what is the good of deliberating whether it is right or wrong? Where is room for the moral law if the natural law overpowers and entirely frustrates it?[4]

Given that this idea is colored entirely by the cause-and-effect mechanisms of physical objects or "dead matter," it is necessarily a *mechanical conception of nature.* It conceives of nature as a giant machine that works purely in terms of the objective relationships between its parts. This places it at odds with what is called a "vitalistic" conception of nature in which it is "animate" from the get-go. Within the vitalistic view (as it is presented here), *nature as a whole is alive.*

When you consider reality and the cosmos and carry things back to the original moment of the birth of our known universe, you ultimately must subscribe to one idea or the other. Either the universe was first an organism or first a machine, one being more primary in the dance between the animate and the inanimate. This is a question of primary importance. Within the context of the whole, which one makes more sense? And if it was in fact an organism first and not a machine, what else might that mean about reality?

THE IDEALIZATION OF NATURE

Essential to the mechanical conception of nature are *idealizations.* We like to think of our science as the pinnacle of pragmatism, that it does not state anything that it cannot demonstrate through empirical evidence. On its face, this seems to say that science is perfectly rational and represents, as truly as we can, that which really happens in the world. Yet there are aspects of our methods and the picture of reality that emerges from it that cast doubt on this perspective.

Much of our science, including key aspects of physics and mathematics, is based on idealizations. To "idealize" in the context of science, refers to the process of simplifying a concept or model by deliberately excluding specific data or details about it. This is done with the aim of creating a more manageable representation of it (that may deviate from the natural world's actual complexity) in order to facilitate scientific analysis or understanding. Idealization often involves taking a phenomenon, stripping it of its real-world complexity, and analyzing it "purely." In a way, it could be said that idealization is a process of *perfection.*

We see this in many places in science, but one of the most obvious is geometry. Geometry is a branch of mathematics that deals with physical shapes, sizes, and spaces, as well as their properties and relationships. It is, alongside arithmetic, one of the oldest branches of mathematics. Euclidean geometry, which had its origins in Euclid's *Elements* written in 300 BC, served as the backbone of geometric calculations for over 2,000 years. One of the foundational aspects of geometry is the manifestation of basic shapes in different physical "dimensions."

All geometry begins with the zero-dimensional *point*. A point has no extension in space. It is considered zero-dimensional because it does not have height, width, or depth. It is simply in one "place" but does not extend itself in any direction. If it began to extend itself in one direction, it would cease being a point, and start being *a line*. A line extends in two directions. It is considered one-dimensional because it has *length* (but does not yet have height or depth). If it had height, it would cease being a line, and start being *a plane*. A plane (like a piece of paper) has extension in four directions. It is considered two-dimensional because it has both *height* and *width*, but does not yet have depth. As soon as it has depth, it becomes three-dimensional, meaning that height, width, and depth are all represented. Each of these physical dimensions—height, width, or depth—extends a geometric figure into a new "space." This is a rudimentary description of "dimensionality" in mathematics.

All of this is well and good. It makes sense and we can picture it in our minds. The problem comes when we try to find it in nature. What the point, line, and plane really are in our mathematics are *idealizations*. A line (sometimes called a "rod" in scientific literature) and a plane (sometimes called a "flat board") that are perfectly straight or perfectly flat are not objects that are observable in nature. In its perfect, idealized form, the line extends infinitely in two directions and does not, at any point in its construction, bend. It is *perfectly* straight and *infinitely* long. Similarly, the board is perfectly flat. It has zero height and extends infinitely in four directions. These are beautiful, precise, perfect images of objects that lend themselves well to the intellect, mathematics, and the imagination. But what is crucial to remember is that, *they are not real*. Austrian-American mathematician Karl Menger describes this:

> The Greeks began the systematic study of objects such as points, lines, planes, polygons, conic sections, and spheres. They discovered how to draw, from a very few assumptions about these objects, an astonishing number of conclusions. Euclid's assumptions (some of which he never stated explicitly) about points, lines, and planes in-

volved a two-fold idealization of the relations between small dots, rigid rods, and flat boards. First, he neglected the extension of the dots as well as the thickness of the rods and boards. Secondly, he assumed the length of the rods to exceed any finite bound.[5]

To reiterate, in zero dimensions this idealization takes on the form of a point with no extension. It has no height, no width, and no depth. In one dimension is a line (or rod) that extends infinitely in two directions. It has length, but no height or depth. In two dimensions there is a plane (or flat board) that extends in four directions. It has width and height, but no depth. Idealizing the point as having no extension in space, the rods and boards as having no thickness, and the length of the rod being without limits takes the idealization to its absolute point. This is useful for understanding the *general* rules and principles in physics.

These idealizations are quite natural to the imagination and to the intellect. In the hands of a disembodied intellect, all forms are made perfect. A square is perfectly square. A circle perfectly round. A line perfectly straight. Why should it "invent" imperfection when it is not there in its own space? This perfection serves to derive the most general principle of a thing, and in that way, it is very useful. But when we move *from the general to the specific*, we begin to get into problems with our idealizations and generalizations. In practice, rather than in theory, these idealizations do not actually exist. They are *idealizations*, not empirical observations.

It is the intellect that works in general rules and guidelines, idealizations and generalizations. The mind and intellect work with general principles. They can think "in the absolute," in objective terms, independent of sensory experience. The senses, on the other hand, what we have established as the other half of the duality inherent to our information processing and cognition of the world, do not work this way. The senses do not work in general principles and objective guidelines. Rather than thinking about the ideal, the senses are working with the *actual* world. Mathematics, which is our primary tool for the quantitative study of nature, is absolutely dependent on these idealized pictures and forms. But the truth is, in the real world, they do not really exist. In reality, *no forms in nature are perfect*. There are no perfect lines, no perfect circles, and no perfectly closed systems.

Any world theory that takes the intellect as its leader and does not sufficiently account for this distinction will necessarily end up with a mechanistic conception of reality. *Given that the mind's nature is to work with things in an objective way in an ideal space, forms will naturally take on their perfect configurations.* Taken to its absolute point, we can imagine the entirety of the cosmos as also perfect. A perfect cosmos with perfect laws in perfect harmony. While this sounds great and is somewhat of a fantasy of the intellect, it is not reflected in observation. There is a factor of *chaos* in the real world that is essential to understanding it. If we do not take this into account, it is because the intellect is our leader. By allowing the intellect to take the lead and create the world picture for us, the idealized form has become the primary one. In this way, by turning the imperfect perfect, our idealizations have actually *distorted* our perception of reality. This takes us to the final and most important idealization of all.

THE FINAL IDEALIZATION

We must not forget that the exclusion of the person of the understander, the subject of cognizance, is also itself an idealization. In the four to five hundred years that we have been doing science in the way we have been doing it, we have forgotten this very simple truth. We have forgotten that considering the world by removing ourselves from it is itself an idealization that is not reflected by observation. As Schrodinger says, it is a "simplifying provisional device," but is not, in the strictest sense, real. Even our most advanced physical science reflected in quantum theory does not permit a purely objective description of nature. The removal of the subject from our concept of nature has contributed greatly to the idea that the cosmos is a machine. Removing ourselves as subjective observers from our concept of nature has given birth to Laplace's demon.

Any physicalist conception of the cosmos, one that imagines a "reality" composed purely of physical objects without the existence of any observers or consciousness, will necessarily be a mechanistic conception of reality. The world of the intellect and mathematics is an *idealized* world. These concepts and methods are based in generalities, not actualities. There is always a factor of irregularity in the real world that eludes the perfection of mathematical precision. Einstein famously said, "As far as the laws of mathematics refer to reality, they are not certain; and as far as they are certain, they do not refer to reality." This is an admonition of the imprecision of the world of forms evident from empirical observation. Nature, while following the laws of mathematical precision as a *baseline*, does not execute on them perfectly. Instead there is always a chaos factor, a manifestation of individual expression which represents a deviation, sometimes major and sometimes minor, from the perfect points and lines of mathematics.

CHAPTER 11

THE ATOM AND THE CELL

S o far our discussion has been mostly confined to physics. When discussing the nature of reality, this is inevitably where our attention turns, given that physics is the science that is closest to the nature of what the world is at the most fundamental level. Still, when we consider the centrality of experience, empiricism, and the subjective observer to reality, there is another scientific field that studies the physical world in a way that is closer to the realm of the living thing, and that is biology.

Biology is the study of the living organism. It encompasses four basic branches: microbiology (focused on tiny organisms like bacteria, viruses, fungi, and algae), botany (focused on plant biology), zoology (focused on the biology of animals), and human biology. There are also several sub-disciplines and specialties beyond these four branches, including genetics, ecology, evolutionary biology, molecular biology, and many others.

In physics, the most basic unit of matter is the atom (even though it also has numerous sub-particles that make it up). In biology, the most basic unit of life is *the cell*. The cell, like the atom, is made up of various sub-components, like a nucleus, mitochondria, and a cell membrane. From the smallest microorganism to the largest mammal, *all living things are made up of cells*. This was discovered sometime around the seventeenth century thanks to the invention of the microscope. Sherrington outlines this development:

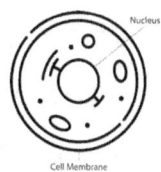

> The microscope, when in the seventeenth century it came, had the cell so to say waiting for discovery. …At a little beyond the limit of unaided eyesight, gland and muscle and indeed all parts of the body resolved themselves into little units of structure which, though characteristic for each, were all of them fundamentally of the same type. To these units Robert Hooke of the Royal Society, an early observer with the microscope, gave the name 'cells'. Practically the whole of all of that part of the animate world which is individually visible to the unaided eye is built up of cells.[1]

With the aid of the newly invented microscope, we were able to get a better picture of the composition of living organisms and the mechanics of organic (versus purely mechanical) systems. Organic systems have a number of peculiar qualities that distinguish them from physical systems. Yet when we analyze these systems at the smallest levels, it is often difficult to tell where "dead matter" ends and "life" begins.

WHERE DOES DEAD MATTER END AND LIFE BEGIN?

One of the most important questions we have about the nature of reality centers around the question of life. Physics makes no opinion about life, neither as an ontological phenomenon nor within the structure of the cosmos. These questions it tends to assign to biology and philosophy. Yet if we are aiming for an understanding of reality as a whole, we can hardly overlook such an important aspect of our existence. In fact, "life" and "existence" are virtually synonymous with each other.

Awareness is the domain of the living organism. The ability to observe phenomena and grow aware of them belongs firmly in the camp of the subject, not the object. Still, if we are keeping the conversation purely to the "outer" domain of physical objects and setting aside our inner experiences, we still run into some trouble. The main question becomes, when observing the physical constituents of a living system, where do we draw the line between what is living and what is non-living? After all, living systems are constantly exchanging energy with their environments. When does the inanimate particle that crosses the boundary into a living system suddenly "come alive?" For this, let us turn to the thoughts of yet another brilliant theoretical physicist of the early 20th century, David Bohm:

As [a] plant is formed, maintained and dissolved by the exchange of matter and energy with its environment, at which point can we say that there is a sharp distinction between what is alive and what is not? Clearly, a molecule of carbon dioxide that crosses a cell boundary into a leaf does not suddenly 'come alive' nor does a molecule of oxygen suddenly 'die' when it is released to the atmosphere. Rather, life itself has to be regarded as belonging in some sense to a totality, including plant and environment.[2]

Although all living things are composed of one or more cells, *all* cells are composed of atoms. To give you an idea of what this looks like, consider the human body. The body is composed of approximately 100 trillion cells, and each one of these cells is alive. If we zoom in on one of these cells and examine its composition, we find that it is made of around 100 trillion atoms. That means that the body is formed of about 100 trillion cells and each individual cell from another 100 trillion atoms! The sheer magnitude of this is staggering, but it serves to put into a clear and visible perspective what the basic building blocks of our human body really are. The laws of physics hold for all of the atoms that form those cells, yet there are other factors involved in the behavior of the cell that are different from that of the atom.

ORGANIZATION, COMPLEXITY, AND EMERGENT PROPERTIES

For example, living systems exhibit a high degree of organization and complexity in relatively small spaces, seeming to follow a different set of "rules" to build themselves up than strictly physical systems. A human gene for example can contain only a few million atoms, yet contains vast amounts of information about an organism's traits and characteristics. But by virtue of what force is it keeping this information organized in such a way? We know that all of the atoms that compose the gene are obeying the laws of physics, but how is it that the greater entities the atoms go on to make up are arranging themselves in such highly regular ways? This highly organized and sophisticated way in which living systems organize themselves at microscopic levels is one of the questions we cannot answer from a purely physicalist perspective.

There is some principle of integration by which these cells (and by extension, the atoms that compose them) organize themselves. British author and academic William MacNeile Dixon puts this eloquently:

...if the atoms or particles, the oxygen, hydrogen and carbon, built up in combination the many million varieties of living organisms which support themselves and reproduce their kind, I should be obliged if you would inform me further whence and how these particles obtained their singular power of cooperating, constructing and organising, of combining together into a unity. The unity of the organism, the co-ordination of its parts, there you have the supreme enigma. The organism involves some principle of integration, some ability certainly not present in the original atoms of carbon or oxygen taken by themselves.[3]

Furthermore, the community of cells that come together to form living systems often exhibit what are called "emergent properties." Emergent properties are phenomena that come from the interactions between individual parts (like atoms, molecules, or cells) but aren't easily predictable or explainable by examining those parts alone. These properties can include complex behaviors, patterns, or functions that become apparent as a result of the organization and interactions between these parts as a whole. A few examples of this could be a flock of birds exhibiting coordinated flight patterns, the complex organization of cells to form tissues, or ecosystems that self-regulate and adapt as species interact and respond collectively to environmental changes.

The point is that in living systems, there are often higher-level phenomena or behaviors that cannot be fully understood by studying individual components by themselves. Instead, they arise from the interplay, organization, and dynamic interactions of these components within the greater system. These emergent properties illustrate a major difference between the natural order inherent to organic versus purely mechanical (inanimate) systems. Organisms tend to organize themselves in more cooperative ways with respect to a greater entity, rather than solely in isolation. *Emergent properties demonstrate that in nature, it is often the case that the whole is more than the sum of its parts.*

In fact, it is this principle of higher order and unity in living systems that is at the heart of the difference between the living and the non-living. While in the realm of the non-living, parts seem to operate on their own without any regard for a greater entity, in the realm of the living, parts always operate in the context of a whole.

THE UNITY OF THE LIVING ORGANISM

One of the most striking aspects of living things is the harmony they demonstrate between their individual and collective aspects. Every living thing is composed of cells yet at the same time, each individual cell contains the genetic code to be able to build up the whole entity again from nothing. When we look at a living thing like a tree for example, we see a creature with a singular identity. But rarely do we consider the fact that this tree is really a combination of uncountable smaller entities that together make it up!

Consider your physical body. This body is made of cells. From the moment of conception until today, this body multiplied itself from one single cell until there were 100 trillion of them. All of these cells operate as their own units with their own lives and jobs, yet each job is also performed in service to the whole organism. You see this same quality over and over in nature, from ant colonies to birds to ecosystems. There is a harmony between individual and collective elements.

As human beings, despite the fact that we experience ourselves as "one" individual entity from the high-level perspective of our ego or personality (and we *are* this one entity), we are *also*, at the same time, a body made of 100 trillion sub-entities, each with their own lives! Schrödinger comments, again invoking Sherrington:

> I find it utterly impossible to form an idea about either how, for example, my own conscious mind (that I feel to be one) should have originated by integration of the consciousnesses of the cells (or some of them) that form my body, or how it should at every moment of my life be, as it were, their resultant. One would think that such a 'commonwealth of cells' as each of us is would be the occasion *par excellence* for mind to exhibit plurality if it were at all able to do so. The expression 'commonwealth' or 'state of cells' is nowadays no longer to be regarded as a metaphor. Listen to Sherrington:

> To declare that, of the component cells that go to make us up, each one is an individual self-centred life is no mere phrase. It is not a mere convenience for descriptive purposes. The cell as a component of the body is not only a visibly demarcated unit but a unit-life centred on itself. It leads its own life... The cell is a unit-life, and our life which in its turn is a unitary life consists utterly of the cell-lives.[4]

While we experience ourselves as "one," the same could be said of each of our cells. Each has a purpose and a function that defines its nature and role within the greater whole. Yet at the same time, all the pieces operate as part of a commonwealth, a community. The greater organism is their common purpose and binds them together.

Another striking factor of this cell commonwealth is that, despite the fact that every cell has its own identity and role within the overall, at the same time, *every cell possesses the entire genetic code to build up the organism from nothing.* Every cell is whole. Self-contained. Just as each of the pieces come together to form the whole structure, the whole structure is also reflected in every one of its parts. In his characteristically elegant prose, MacNeile Dixon comments:

> The germ cell [stem cell] is a unity and does not become specialised for the production of the heart or lungs, or any other part of the body till it has attained a certain maturity. If at an early stage it be divided or subjected to pressure, or even if a portion be removed, the germ retains all its powers. It possesses the astonishing faculty of providing any necessary organ out of any part of itself. Utterly unlike any machine, the cells, too, in living things can act for each other, and work together for a common purpose. This co-operation of parts is everywhere present in natural organisms.[5]

When we examine the difference between the animate and the inanimate, it is this propensity for each part of a living thing to demonstrate both an individual and a communal identity at the same time that is one of its most important features. Living organisms are unities, and at the same time, pluralities. Nature has built a failsafe into this relationship in that the whole is always represented in the parts (in the code at the center of the cell), and the parts are also represented in the whole (in the cells that make it up). This interrelationship between the whole and parts represents a harmony to the natural order evident in all living systems.

THE WILL

We must be absolutely clear in that, despite the fact that this branch of science studies "life" and not the inanimate objects of physics, it still does not deal with the reality of the subjective observer. Biology is still concerned with life at the physical level and studies the behavior and composition of living cells in much the same way that physics studies atoms. Mostly, it is concerned with function, composition, and behavior. Even still, certain aspects of living organisms fundamentally set them apart from the universe of the inanimate. One of the most visible of these is *the will.*

Sometimes called "volition," the will pertains to the aspect of the living organism that chooses. An organism, versus a "dead" physical object, is *self*-willed, *self*-directed, *self*-centered, so to speak. This will allows it to direct its actions in one way or another. Within a living system, each entity may choose to go about things in its own way, whereas with physical objects, entities always follow physical law and cannot "choose" to deviate.

Like purpose, the will is not a phenomenon that can be witnessed or measured by physical instruments. It is not visible from the external perspective looking at inanimate objects, but is visible and discernible from the internal perspective of the observer looking out. It speaks to the ability of an observer to affect their material environment through the focus of their attention. If physical law is all we consider, this of course poses a problem. Even Bohr was conscious of this at some level:

> In any attempt to pursue the enquiry we must, of course, be prepared to meet increasing difficulties at every step, and it is suggestive that the simple concepts of physical science lost their immediate applicability to an ever higher degree the more we approach the features of living organisms related to the characteristics of our mind.

> To illustrate the argument, we may briefly refer to the old problem of free will. From what has already been said it is evident that the word volition is indispensable to an exhaustive description of psychical phenomena, but the problem is how far we can speak about freedom to act according to our possibilities. As long as unrestricted deterministic views are taken, the idea of such freedom is of course excluded.[6]

It should be noted that the purely deterministic model of reality put forward by science for hundreds of years essentially died with the more advanced probabilistic understandings that arose with quantum mechanics. We realized that uncertainty was a feature baked into nature itself and that our "predictions" were only valid up to a certain measure of uncertainty. Nature was not necessarily strictly determined at the outset, but was inherently probabilistic. Yet admitting that the physical universe is not strictly deterministic is still a far cry from admitting that we as observers have anything to do with that indeterminism. Aside from "the observer effect" (which still does not have universal acceptance in scientific circles), we remain content to leave these types of questions outside of the analysis of our observable environment. The ability for living organisms to some extent direct their actions and affect their bodies at a cellular level are considered "interdisciplinary fields." The idea that such a thing should be *fundamental to understanding reality itself* is rarely, if at all, considered.

What we sidestep with these omissions is the seemingly reversible interaction between the "I" and the body. From the practical perspective of the observer, this has two aspects to it. On the one hand is the scientific aspect which aims to understand the cause and effect of it. By what *mechanism* do our internal activities affect our physical bodies? What are the *causal laws*, the laws of nature, which give order to the interaction between our conscious minds and physical matter? The second half is the apparent sense of responsibility that comes from ownership of our actions. If everything I do is simply the result of chemical interactions in my body, then does that mean my *body* is making all the bad decisions in my life that lead to sometimes very negative consequences? Have "I" no influence in the matter? Again, this seems to make a lot of sense from a perspective of the cosmos which holds physical matter to be primary in all things. But from the perspective of the one who is having the experience looking outward, from the perspective of the subject, not so much. Schrödinger continues:

> ...Since my mental life is obviously bound up very closely with the physiological goings on in my body, more especially in my brain, then, if the latter are strictly and uniquely determined by physical and chemical natural laws, what about my inalienable feeling that I take decisions to act in this or that way, what about my feeling responsibility for the decision I actually do take? Is not everything I do mechanically determined in advance by the material state of affairs in my brain, including modifications caused by external bodies, and is not my feeling of liberty and responsibility deceptive?[7]

Something is missing here. Clearly our current models cannot account for this. What is required is a new way of thinking—a framework that can explain the causal laws that give order to the reversible interaction between the "I," the subjective observer, and their material environment. Furthermore, the will opens up the question of how individual behaviors are motivated by factors beyond the strict causality of physics, including action for a purpose.

LIFE AND PURPOSE

When it comes to the biological organism, standing opposite to the inanimate matter of physics, we again return to the issue of purpose. Sometimes called "teleological" or "finalistic" arguments, standing in opposition to that which is termed "mechanistic," purpose has to do with the *reason* things happen in a particular way. Essentially, *to what end*, beyond the simple cause and effect of physics.

When entering into the discussion about the differences between living organisms and dead matter, the idea of purpose ultimately comes up, because while this kind of "having a purpose" may be absent from objects, the notion presents itself clearly in the behavior of living things. Living organisms are not always reacting to pure mechanism but sometimes act with purpose for an end—like for food, pleasure, or survival. In this way, their movements and behaviors are not always mechanically motivated, but purposeful. We return to Bohr:

> Without attempting any exhaustive definition of organic life, we may say that a living organism is characterized by its integrity and adaptability, which implies that a description of the internal functions of an organism and its reaction to external stimuli often requires the word purposeful, which is foreign to physics and chemistry.[8]

There are many reasons or "purposes" for which living things can live: food, reproduction, survival. But chief among them is one which pierces to the essence of our discussion. It is a ubiquitous force observable in nature that manifests itself everywhere, from the single-celled organism to the multifaceted human being—and that is *the will to live*.

INNER STABILITY AND THE WILL-TO-LIVE

One of the most potent forces to manifest itself in the experience of the subjective observer and visible across all manifestations of life from the simplest to the most complex organism, is the will to live. What we observe about life is that it *strives* to continue. Everything about the living organism demonstrates this "urge" to maintain itself, even in the face of sometimes harsh or demanding circumstances. This is one of the most stark distinctions between the atomic and the cellular, the living and the non-living. We return to Dixon:

The will-to-live is ubiquitous, universal, insistent. Nature advertises it, all existence manifests it, life in every creature gives it the clearest utterance, and well we know it in ourselves. There the hounds of this desire to be alive and remain alive are in full cry. So profound and pervasive is the instigation of this instinct, upon which all else appears to rest, that we might well conclude with Schopenhauer that it is more fundamental than thought or mind, and gave birth to the whole creation. For we cannot dig deeper to find a surer foundation.[9]

This is sometimes simply referred to as "biological drive" or "survival instinct." It represents the innate, instinctual desire of living beings to ensure their own continuance. It motivates various behaviors and physiological processes aimed at sustaining life, such as seeking food, avoiding danger, or reproducing. In simple terms, it is the fundamental urge to stay alive that is hardwired in all living creatures.

Wherever this urge might come from, this propensity for living systems to maintain their inner integrity, even in the face of external circumstances that are sometimes harsh and inhospitable, is one of the defining characteristics of life. There is a sort of *inner* stability to the living organism. This "inner space" is what attempts to maintain itself, distinguished from the "outer" space of its environment and other organisms. This incredible inner stability or "inner homeostasis" demonstrated by living organisms is perhaps the most striking feature that distinguishes it from inanimate objects in physical terms. Austrian physicist and philosopher Ernst Mach—many of whose ideas influenced Einstein in his early years—summarized this nicely:

Every organism together with its parts is subject to the laws of physics. Hence the legitimate attempt gradually to conceive of an organism as something physical, and to establish the consideration of it in a "causal" point of view as alone valid. But whenever we try to do this we are always brought face to face with the peculiar characteristics of the organic, for which no analogy can be found in the physical phenomena of "lifeless" nature, so far as they have been investigated at present. Every organism is a system that is able to maintain its peculiar properties,—its chemical composition, its temperature and so forth,—in the face of external influences, and which manifests a state of dynamic equilibrium of considerable stability.[10]

Life is characterized by this inner stability. How such an incredible force demonstrates itself in physical systems is beyond the explanatory powers of both physics and biology, which aim to understand the physical behaviors of these systems. To understand it, we must return to that characteristic of life so very close to its essence and nature—*unity*. Life always manifests itself as a unity. It demonstrates this principle at every turn. There is the unity of the parts, each piece and part working together as a commonwealth in favor of the whole. There is the unity of the whole, demonstrated in its entirety in the seed at the heart of every cell. There is the inner homeostasis, stability, and inward harmony demonstrated in the physical composition of every living thing. There is its utter loyalty to itself, the fundamental urge at its core to protect its own integrity and continue on.

When we consider whether the whole universe is really an organism or a machine, we might consider this unity. Life exhibits qualities that bear great resemblance to the behaviors visible when our universe is observed as a whole, rather than in its parts. Could it be that the unity of our experience, the unity of the cosmos, and the unity of physics, are actually all intertwined? The organic (animate) and the mechanical (inanimate), rather than being two disparate aspects of the cosmos, may actually be bound up with each other. In other words, they may *both* be true. Rather than being contradictory, they may be *complementary*. Bohr says:

> ...It must be realized that the attitudes termed mechanistic and finalistic are not contradictory points of view, but rather exhibit a complementary relationship which is connected with our position as observers of nature.[11]

If such a thing were true, that would mean that to understand the nature of the cosmos as a whole—the nature of reality—we must take both into account in whatever ontological model we develop. To account for them both, their individual natures and their relationship with each other, we create the opportunity for a system that can truly speak to the whole. Not only to the nature of the objective universe as it exists in general terms, but also to the observer, their individual experience, and its role within the order of the cosmos.

CHAPTER 12

DNA AND IDENTITY

We established in the previous chapter that the basic unit of life is the cell. While it's true that this is the smallest unit from which biological organisms construct themselves, there is actually another unit contained within the cell that could be considered the *true* seed of life, and that is DNA.

Cells fall into one of two categories: prokaryotes and eukaryotes. There are only two types of prokaryotes: bacteria (which are extremely common) and "archaea" (which used to be classified as bacteria but have since been put in a separate category and are more rare). Prokaryotes are unicellular (meaning they only have one cell). As opposed to eukaryotes, which have clearly defined sections or membranes, inside prokaryotic cells there is a sort of "open space." Instead of sectioned-off (membrane-bound) structures inside the cell, they have "regions." In the central region of the prokaryotic cell we find its genetic material, typically in the form of a circular band of DNA. This is a basic picture of prokaryotes.

All living things other than bacteria and archaea are composed of eukaryotes. This includes animals, plants, fungi, and "protists" (a group of organisms that includes algae). Eukaryotic cells are generally larger and more complex than prokaryotes, therefore, while prokaryotes vastly outnumber eukaryotes in nature, in terms of biomass, they are closer to equal. One of the main differences between these two types of cells is the presence of a clearly defined nucleus. In eukaryotic cells, the genetic material is contained within the nucleus, while in prokaryotes (which have no nucleus) it typically exists in the form of circular DNA free-floating toward the center of the cell.

DNA stands for deoxyribonucleic acid (its full chemical name). It contains the genetic material of the organism and is typically found in the central region of the cell. Just as atoms contain a nucleus at their center, in the form of protons and neutrons bound together and which determine their "identity" within the Periodic Table of the Elements, eukaryotic cells also contain a nucleus at their center which, by way of their DNA, similarly establish their identity.

DNA is a code. While it often takes a circular form in the more primitive prokaryotes, in all other living things it takes the form of a "double-helix." This double-helix structure is composed of two strands that spiral anti-parallel to each other (meaning in opposite directions) and are connected by little bridges. These bridges consist of substances known as "nucleotide pairs," one on either strand connecting with the other in the middle. There are four of them, and each has a specific one it pairs with: adenine (A) which always pairs with thymine (T), and guanine (G) which always pairs with cytosine (C). These four nucleotides are like letters of a genetic alphabet that spell out a genetic "language" or code. The double helix structure contains the entire code of the living organism, all the way from the general characteristics that make up its physical body down to the individual variations that can occur within its own genetic line.

DNA is unique to every living entity. In this way it can be thought of as a kind of "biological fingerprint." It is the closest to a "seed" of life we have in physical, three-dimensional terms. It contains within it the instructions that facilitate the organization of the atoms and molecules that go on to build up the living organism from its environment. *All living things possess DNA.* Just as all living things are made of cells, all cells contain DNA. In relatively rare cases a cell will ditch its DNA after it "grows up" and becomes specialized (like red blood cells, for example), but at least from birth, *all cells contain the entire genetic code of the whole organism they represent.*

INDIVIDUALITY IN NATURE

DNA speaks to one of the most fascinating and important aspects of biological life: *individual identity*. In the world of physics, there really is no such thing as individual identity. While there can be specific situations and events, physics is always after the general and universal, not the individual and specific. This is one of the primary reasons it is ill-equipped to deal with the phenomenon of life. Life is everywhere marked by the prominence of the individual. Every tree, flower, river, or mountain has an atomic configuration that is a little bit different. *No two living organisms are exactly the same.* This is one of the great miracles of nature. The ability to be something unique—not that which is around you but something all your own—is a mark of all life.

What makes this individuality tricky to pin down is where it arises within the natural order. The question is, *why* are no two living things exactly the same? What gives rise to this individuality in natural systems? And how does this fit into the cosmic order?

As we discussed, one of the elements that sets living things apart from inanimate objects is *agency*. Supposedly a living thing can *choose* to act in one way or another, respond to stimuli, or to adapt itself, taking us full circle back to the idea of the will. This will introduces a "chaos" (irregular, unpredictable) factor into the ordered physical and biological systems of nature. But, in truth, it is not even agency that is at the core of what produces individuality in living organisms.

There is a phenomenon even closer to the essence of it that is at the heart of what it means to be alive, and that is *experience*. To be able to experience the world in some way, even if only through extremely simple sensations like pressure or temperature, is what truly marks the difference between life and inanimate matter. Because every living organism experiences reality through its own unique and relative frame of reference, nature ensures that *all experience is unique.*

The uniqueness of experience is at the very essence of subjectivity. It ensures diversity and individuality in all living things. Because every experience pertains to a subject—an entity or organism with a unique frame of reference—no two experiences are identical. As these individual experiences accumulate within the living organism, this aspect of individuality only grows more pronounced.

It is this individual aspect of nature—that which occurs because of experience—that belongs more to subjectivity. The collective or general aspects belong more to objectivity. The objective works in general rules that should be applicable in all or most cases. The subjective works in absolute specificity, for one thing and nothing else. This is hinting at another fundamental duality in the cosmos, visible in physical systems of nature as much as it is within our own experience. It is a duality that lines up nicely with subject-object, subjective-objective, and sensory-intellectual dichotomies we have already set up—and that is the duality between *the individual* and *the collective*. This duality is reflected everywhere in nature and in biological systems. Yet, from within a lens of pure physics, it is not visible. This is one of the unavoidable shortcomings of purely physical models that aim to reconstruct reality only from material components.

This leads us to another important question. If every living thing is having an individual experience, owed to its senses and unique frame of reference, how does this experience affect physical systems?

THE EFFECT OF SUBJECTIVE EXPERIENCE ON PHYSICAL MATTER

One of the clearest ways physical biology is affected by subjective experience is *memory*. Memory is the way by which experiences can persist through time and influence behavior. As we have already established, while in physical systems behavior is dictated solely by physical laws, in biological systems, behavior is also driven by many other factors.

The memory of experience can be recorded in the body in a number of ways. Two of the most important of these are called "neuroplasticity" and "epigenetics." Neuroplasticity refers to the ability of the brain to change. The brain can reorganize itself by forming new neural connections throughout its life, allowing neurons (nerve cells) to adjust their activities in response to new situations or changes in the environment. Essentially, neural networks in the brain can change over time through growth and reorganization.

A more specific form of neuroplasticity known as "synaptic plasticity" is believed to be the way the body stores memories. A synapse is the area where information transfer occurs between neurons, allowing them to form networks and connections. Synaptic plasticity refers to the ability of synapses to change their strength over time. Repeated activation of a synapse can lead to something called "long-term potentiation (LTP)," strengthening the connection, while decreased use can lead to "long-term depression (LTD)," weakening the connection. Memory is believed to be recorded in the synaptic pathways of these neural networks.

The exact mechanisms by which memories are encoded in synaptic connections, how they are represented in neural networks, and how they are retrieved, are still not well understood. What we do know is that new memories are handled mainly by the hippocampus, an S-shaped region of the deep brain, and long-term memories are sometimes consolidated and moved to the cortex (the brain's outer layer). Strong emotional experiences can activate the amygdala which seems to enhance consolidation, making them more stable and easier to recall later on. This means that experiences accompanied by strong emotions, whether positive or negative, are often remembered more vividly and accurately than others.

Besides the neural storage of memories, the body also has other forms of memory. This includes the immune system's ability to remember previously encountered pathogens, a phenomenon known as immunological memory. Unlike neurological memory which works with synaptic pathways, this type of memory is the result of specific configurations of receptors encoded into a cell's surface which are created by biochemical processes that alter the cell's structure. These specialized receptors allow these "memory" cells to recognize and rapidly respond to pathogens they have encountered before. This is the main principle behind vaccinations, where exposure to a vaccine prompts the creation of memory cells, enabling a quicker and more effective response if the pathogen is reencountered later. While this form of cellular memory is crucial for immunity, it operates through mechanisms different from the synaptic plasticity and neural networks involved in the storage of memories in the brain.

This incredible adaptability of the brain and body in response to our internal experiences and physical environment highlights the enormous variety of ways our physical form can shift and change with us as we move through life. The continuous interplay between the "I" and our physical form, the dance between our thoughts, feelings, and experiences and our brain's structure demonstrates the profound influence of subjective experience on biology, molding our neural pathways and synaptic connections. But neuroplasticity is not the only or even the most significant way that our subjectivity, the experience of the living organism, affects physical matter. The greatest among these could be considered epigenetics.

The field known as "epigenetics" deals with the way the individual experiences of living organisms interact with their genetics. It describes the way behaviors and experiences can cause physical changes in gene "expression" (how our genes are "read" or "used") without actually changing the underlying DNA. In simple terms, the organism's experiences and certain environmental factors can turn genes on or off, so while the underlying DNA doesn't change, the way the body reads it does. This can cause an emphasis or deemphasis of certain traits or characteristics which are available in the DNA, but which may or may not be active.

It is something like a painter. The various colors they have available on their palette are the genes in the gene pool available for use. But the gene expression, the "epigenetics," are the colors that are "picked out" and actually used and combined on the canvas. We can also picture it as a guitarist who has several strings and chords they can use on their instrument to make music. The strings and chords are the gene pool available in the DNA, but the ones they actually choose or "activate" on the guitar to give life to their composition would be the gene *expression*, the epigenetics.

Gene expression is greatly influenced by our experiences and choices. It can change according to things like environment, mentality, diet, lifestyle, and emotional states. What is important to realize is that *epigenetic changes are happening in real-time*. In other words, our choices and experiences are literally changing our physical bodies, both neurologically and genetically, with more extreme or consistent experiences having a more pronounced effect on our epigenetic expression. Of course not every experience or choice will directly result in a detectable or significant epigenetic modification, as some might manifest over an extended period, but the point remains. Our genetic expression directly responds to our internal, subjective states.

It is important to note that these epigenetic changes can be passed down to our offspring, bearing a direct influence on evolution. They can sometimes provide an advantage to subsequent generations by creating an extra layer of adaptability in response to environmental pressures or stresses which can enhance survival and reproductive success. This complements slower, more permanent changes that occur in the DNA over vaster periods through factors like mutation and natural selection. The interaction of experience-induced epigenetic changes along with inherited genetic material in DNA crafts a dynamic conversation between the living organism and its physical environment, shaping a pathway for evolution in biological systems.

LEARNING, IDENTITY, AND EVOLUTION

Memory is an incredible feature of living things that makes them dynamic and adaptable. It ensures that an organism can use information from past experiences to inform its future decisions, whether to ensure its survival or to improve its quality of life. That experience can affect behavior through memory is a core quality of living things that is absent from inorganic matter. We cannot say that an oxygen atom "experienced" separation from hydrogen in a chemical reaction and therefore later avoided hydrogen. This is one of the main differences between organic and purely mechanical systems. Living systems can *remember*. Ernst Mach comments:

> What is memory? A psychical [mental] event leaves psychical traces behind it, but it also leaves physical traces. Physically, as well as psychically, a child that has been burnt, or stung by a wasp, behaves in quite a different way from a child that has not had this experience. For the psychical and the physical are different only according to the way in which they are regarded. Nevertheless it is extremely difficult to discover in the physical phenomena of the inorganic world characteristics having any affinity to memory.
>
> In the physics of the inorganic world everything seems to be determined by the circumstances of the moment, and the past seems to be entirely without any influence. The oscillations of a pendulum are equal, whether it is performing its first oscillation or whether 1000 others have already taken place. Hydrogen combines with chlorine in the same way, no matter whether it was previously in combination with bromine or with iodine.[1]

There is an implication in this that is at the heart of the nature of life and subjective experience. *We can learn.* When a gazelle wanders into the wrong part of the savannah where lions are active and loses one of its offspring, the experience of that loss is likely to change its behavior in the future. When we burn ourselves over and over on a hot stove, it is likely to cause us to approach that stove with more caution in the future. Our experiences are not just static things that come and go with no material influence on our behavior or choices. And this is the point. *Experience affects behavior.*

This is to say that what we learn from our experiences goes on to affect our choices. Our capacity for *agency*, that we have a will which allows us to direct to some extent our actions, ensures that experience will cause change in the physical world. This is where we begin to get to the core of the interaction between the subjective and the objective.

There are general laws in the world. There are patterns that guide the hand of the movements of all bodies and forms. This is irrefutable. But as it stands, it is not enough alone to explain everything. We must also take into account the reality of the individual. The effect of past circumstances on present behavior. Even single-celled organisms can learn, adapt, and change, so this phenomenon is not limited to the domain of man and higher life either. It is present in nature at all levels. The will of living things that allows them to some extent to direct their actions, and the inherent uniqueness of experience brought about by the singular frame of reference belonging to every living entity, ensures change, dynamism, and individuation at all levels of the living world.

These three elements, experience, learning, and behavioral change, are what make evolution possible. Evolution is the process by which living (and by extension physical) systems change over time. Our DNA, which is the living code that informs physical matter as to how to build up an organism, grows and changes through time by way of the choices and experiences of the living entities that make it up. This happens through mutations which can occur for a variety of reasons, as well as the effect of gene expression and epigenetics on natural selection.

DNA lives at the heart of all life. At the center of every cell. You could say that the entirety of the organism is contained at the center of its seed. When the organism that began as a seed matures, flowers, and produces seeds again, the cycle is complete. But this time, the genetics are not the same. They have been slightly modified by the experiences of the organism that came before them.

This ensures that the information belonging to subjective experience is not lost, but is recorded and directs change. It would be an utter waste if the depth, variety, and significance of individual experience and all that is learned and discerned from it were just "thrown out" the moment an organism died. *It is an observable quality of nature that it does not waste.* By recording the most pertinent information of an organism's lifetime in its genetic code, it ensures that there is *continuity*. It is this continuity, and the preservation of elements of subjective experience within a code, that facilitates the process of evolution.

It is only because of learning and the record of change within our genetics that evolution is possible. If everything were pure physical mechanism, evolution would not exist. Under that thinking, any sort of "change" to behavior would not be possible because behavior would be strictly directed by physical laws. Therefore it could only change if the laws of physics themselves changed. Subjective experience, learning, and behavioral change must be accounted for. If we are to arrive at an integral and consistent ontological framework, these observable nuances of our biological reality must be reconciled with the physical one to come to an understanding that is whole.

PART IV

THE EAST

& THE WEST

CHAPTER 13

THE WHOLE AND THE PARTS

By now we have come to understand some of the wider questions we have as spectators and participants in this great dance of life. Life has qualities and attributes beyond those which appear in a purely mechanistic and quantitative view of reality. These attributes must somehow be included in our total accounting of things and integrated with the understanding of physical law we have developed over the last half millennia.

Interestingly, this quantitative, object-oriented approach to nature is actually quite particular to the West. In the East, questions about the ultimate reality often took a more "spiritual" turn. Models of the nature of reality often included and even emphasized qualitative, conscious aspects of our existence. In the East, the "experiencer" was frequently at the core of the total order of reality, its cause and its purpose.

THE TOTAL ORDER OF REALITY

We established that an understanding of the objective processes of the physical world allow us to build technology. By understanding physical patterns, behaviors, and laws, we can act on nature in intentional ways to build tools and machines. But what happens when we understand the patterns and behaviors—the causal mechanisms—that are behind the observer's inner life? What happens when we understand the objective processes, the *laws* behind subjective experience and the observer's inter-action with the material world (and other observers)? At the least, it would allow us to facilitate intentional changes to our conscious experience. In very favorable circumstances, it could allow us to build new machines that act on the causal mechanisms between subject and object. At its best, it could pave a path to enlightenment.

"Enlightenment" is sometimes described as an experience of direct contact with one's true self or the fundamental nature of existence. It is characterized by feelings of joy, inner peace, liberation from suffering, clarity, and unity with the cosmos. What we know is that enlightenment is not really a question of material objects, but pertains more to *experience*. Experience, being refractory to measurement, and therefore also to mathematical treatment, is outside the scope of our current cosmology. This is why metaphysics, as we've defined it in this work, is so important. If science is aiming for a total understanding of the mechanics of objects and material reality, then metaphysics adds to this the reality of the conscious subject and the immaterial reality of their experience, to come up with a cosmology which can account for the total order of reality, material and immaterial, conscious and physical.

If we consider the broader implications of this, metaphysics, in its true form, is a science of energy. It is a means to understanding cosmic ("conscious") laws, and in its most ideal expression, a path to enlighten-ment. By seeking an understanding of the total order of reality, *all* natural laws, the way things work materially, but also those laws pertinent to the subjective observer and their experience of the world, it aims to recon-struct, as best it can, a model that includes all the entities in existence and the relationships between them. This is a true and total metaphysical ontology. A system for the total order of reality and all the entities that exist within it. Nevertheless, to discern this natural order and understand its mechanics, we must ask a different set of questions.

The question of an ultimate understanding as it relates to the lived experience of the world, rather than simply to the so-called "objective" viewpoint of nature which focuses on the object and excludes the observer, necessitates a different methodology. It is precisely because of the necessity of a different methodology that the worlds of science and spirituality have never really been able to play together. So much of what happens in the supposedly "spiritual" world cannot be measured physically, and is therefore outside of the scope of scientific analysis.

Nevertheless, as we have said many times already, if we are taking a holistic account of nature, *all* that we can witness—quantitative and qualitative, conscious and material—then the whole order of things, all entities that are observable, must come together in our model for it to be considered whole. This ensures that the psychological framework at the base of our thinking, the network of connected logic upon which our interpretation of reality is based, will not come from a worldview that is fragmented or fractured.

QUANTITATIVE ANALYSIS AND "THE PARTS"

Part of this has to do with the way our inquiry into nature has evolved. Early on in our history, taking reality as a whole was a given. We did not question whether this was the right or wrong way to look at things. But beginning with the Scientific Revolution and the methods of analysis that gained dominance from it, breaking things down into smaller pieces became the name of the game.

Analysis is a word with the Greek root "lysis," which implies a breaking apart or dissolution of things. This is similar to the way a mechanic or an engineer sees things. A mechanic must understand the "pieces" and the relationships between them, then put them together in a way that creates a synthetic whole. It is important to realize that this process of "analysis" or "breaking apart" can happen with more than just objects—it can also happen in thought. Bohm comments:

> ... it is useful first to note that the word 'analysis' has the Greek root 'lysis', which is also the root of the English 'loosen' and which means 'to break up or dissolve'. Thus, a chemist can break up a compound into its basic elementary constituents, and then he can put these constituents back together again, and thus *synthesize* the compound. The words 'analysis' and 'synthesis' have, however, come to refer not merely to actual physical or chemical operations with *things*, but also to similar operations carried out in *thought*. Thus, it may be said that classical physics is expressed in terms of a *conceptual analysis* of the world into constituent parts (such as atoms or elementary particles) which are then conceptually put back together to 'synthesize' a total system, by considering the interactions of these parts.[1]

From the time the scientific method took over as our primary way of assessing the world and our collective outlook began to be affected by it, our worldview started to become more and more fragmentary. The holistic viewpoint, which could be considered the most natural and common sense way to look at the world, was slowly replaced by an emphasis on seeing things in their parts. d'Espagnat describes this:

> Primitive societies have a vision that might rather be termed "holistic." But it is quite unquestionable that, from the times of Galileo and Descartes on, the growth of science was largely grounded on a rejection of the said vision. In fact, it greatly relied on its exact opposite, that is, on the principle—explicitly stated by Descartes—according to which problems and the objects they deal with have to be divided by thought into simple elements before the whole is reconstructed. And indeed, throughout the last centuries the success of this method was so impressive that, quite normally, we came to consider it an indication that reality itself is essentially composed of parts.[2]

Part of this has to do with the nature of quantitative analysis, one of the cornerstones of the scientific method. In the duality between the whole and the parts, it is the parts that lend themselves well to the quantitative. When we take a whole and divide it into smaller pieces, we get a discrete set of units we can count. When we can count, we can quantify. We can analyze. The data suddenly lends itself to mathematical treatment. *Even physical measurement is a breaking apart of a whole entity into smaller pieces that can then be quantified and analyzed.*

But when we begin speaking of "the whole," we are no longer talking about pieces and parts. We are speaking of an entity in its entirety. When we begin to speak of something in wholeness, we open the door to speak of its *quality*, rather than simply its *composition*. *Quality is revealed in wholeness, and composition in the parts.*

Let's take a qualitative phenomenon like beauty, for example. Beauty is not matter. There is no clear physical element it corresponds to which we can then isolate and carry into other systems. Beauty cannot be measured with a ruler. *We cannot dissect beauty intellectually and still perceive it.* Like other qualities, beauty is perceived in wholeness. It is a phenomenon that is more than the sum of its parts. Indian philosopher and mystic Osho comments:

> Intellect divides, cuts into pieces to understand a thing. Science is based on intellect, dissection, division, analysis. Intelligence joins things together, makes a whole out of parts -- because this is one of the greatest understandings: that the part exists through the whole,

not vice versa. And the whole is not just the sum of the parts, it is more than the sum.

You can have a rose flower, and you can go to a scientist, to a logician. You can ask him, "I want to understand this rose flower"; what will he do? He will dissect it, he will separate all the elements that are making it a flower. When you go next you will find the flower gone. Instead of the flower there will be a few labeled bottles. The elements have been separated—there will not be any bottle on which will be the label 'beauty'.

Beauty is not matter and beauty does not belong to parts. Once you dissect a flower, once the wholeness of the flower is gone, beauty is also gone. Beauty belongs to the whole, it is the grace that comes to the whole. It is more than the sum.[3]

It has been said that when we begin to speak of life in its parts, it loses its miracle. Much of this has to do with the fact that all the things that make life beautiful belong more to our senses than to our intellect. Beauty is a feeling, a perception, an *experience*. It is fundamentally qualitative. Experience, as we have established, is connected to our capacity to feel. It is linked to our senses. While our senses can perceive quantity, they are attuned to quality. They are meant to perceive things in their wholeness.

The intellect on the other hand, the reasoning mind, is inclined to break everything apart. It dissolves and divides. The concept of order—which is a reasoned and systematic composition of parts—pertains more to the intellect. The intellect is inclined toward composition and analysis. In the relationship between these two sides of the cosmic order, we have identified yet another duality.

The subject and the object, the subjective and the objective, the sensory and the intellectual, the intuitive and the rational, the qualitative and the quantitative, the whole and the parts, *these are speaking to one and the same duality. A duality that is baked into the fabric of reality itself.*

THE CAUSAL ORDER - AN UNFOLDING OF PARTS

As soon as a whole is split into parts, there is also born in the same instant a relationship between cause and effect. The whole is necessarily the *cause* of the parts, because it is their origin. The parts follow from this primary cause because they are always dependent on a whole within which to exist. Parts *imply* a whole. Logically, therefore, in the duality between the whole and the parts, *the whole is primary*. Primary and secondary are foundational concepts in any ontological system because they are related to the concept of order.

Causal order poses the question, in the chain of cause and effect, what is primary and which is secondary? Whatever comes first is the progenitor for those which follow, since they arise from it. The existence of a causal order comes from the occurrence of time. Since we live in a world where there is a sequence to things—a before and an after—it is meaningful to consider in what order things occur. Among each of the elements we can identify, in what order do they occur in the order of time?

Understanding causal order is like discerning the seed (the cause) that brings forth a tree (the effect) in a garden with diverse flora. When we know which seed leads to which tree, we can comprehend the garden's makeup and predict which trees will grow where. It's a tool that enables us to untangle the web of interrelated events in the world around us, whether it's understanding how a change in weather patterns affects local ecosystems or how our daily habits will later influence our long-term health. In essence, understanding causal order is the key to unlocking the intricate dance of cause and effect in the world, allowing us to make sense of the past and therefore navigate the present.

If you remember from our discussion of organic and inorganic systems, in organic systems, the whole is always represented in the parts and the parts in the whole. This is best exemplified by the relationship between the seed and the organism. A seed represents at the same time the grown and the ungrown, the manifest, and the potential to manifest. The DNA within its heart *is* the tree, in all its glory, not yet manifest. When the seed has been planted and given the chance to grow up to maturity, it produces flowers which then drop more seeds, ensuring the cycle continues. The seed is present at both the beginning and the end of the creation process, and contained within it is the genetic code which represents the entirety of its "self" in physical form.

In inorganic systems, this is not the case. There is no "unfolding" of a whole from within a seed (a part). In a mechanistic worldview, each piece and part of the system is essentially separate from each other and brought together "externally" to create a whole. When we consider the opposite situation well represented by living organisms, each of the parts are instead contained within the whole from the beginning, and are then "unfolded," like the body from a seed or an oak tree from an acorn. In this way, the parts are not separately existing entities, but are actually part of the same structure. An impression of the whole is present in each of the parts, which are self-contained extensions of the greater body. Sherrington says:

> . . .The harmony of the whole is not merely built out of its parts but is impressed on the parts by the whole. An individuality whose whole, as luminously said by Coleridge, is presupposed by all its parts.[4]

What is implied by this—visible in organic systems, but missing from inorganic ones—is *interdependence*. Parts growing organically from a whole imply that each part, in addition to being independent on some level, is also *interdependent* with each of the other parts it shares a structure with. When one part of the organism affects another, it is at the same time affecting its "self," because they both belong to the same structure. In a mechanistic view of the world where each of the parts are separately existent and come together "externally," there is no interdependence between the parts. Bohm continues:

> Let us first consider the mechanistic order. ...The principal feature of this order is that the world is regarded as constituted of entities which are *outside of each other*, in the sense that they exist independently in different regions of space (and time) and interact through forces that do not bring about any changes in their essential natures. The machine gives a typical illustration of such a system of order. Each part is formed (e.g. by stamping or casting) independently of the others, and interacts with the other parts only through some kind of external contact. By contrast, in a living organism, for example, each part grows in the context of the whole, so that it does not exist independently, nor can it be said that it merely 'interacts' with the others, without itself being essentially affected in this relationship.[5]

This takes us back to the question of whether the universe was living from the get-go, or began "dead," and later manifested life. The inherent differences between organic and inorganic systems especially with respect to the relationship between the whole and the parts has important implications for the nature of the universe, depending on which came first in the causal order. It is especially this principle of unfolding, that each part of an organic system has an imprint of the whole within it and operates with a sense of interdependence and intrinsic connection with every other piece that makes this important. For the parts to share an *internal* relationship would have consequences on their behavior. When it comes to the nature of reality, understanding the causal order is important. Was the universe originally *alive*, and a cascade of complex mechanical elements followed, allowing life novel avenues of expression? Or did it begin as a dead machine, eventually manifesting the living organism?

At the very least, what we can infer from logic is that the universe did not begin in parts. The word "part" is related to "partial," meaning not whole, and hence implies a greater entity. All things must necessarily begin in wholeness. By the process of analysis, we can take a whole apart to study its composition, but it is imperative to remember that these compositional elements belong to a greater entity. When we lose sight of this, it is easy to begin considering them as separately existing and forget about the greater context within which they exist. In an ant colony for example, each ant has a job. Some are foragers, defenders, a queen, etc. But while each ant possesses its own autonomy and acts independently on some level, its actions must always be considered within the context of the whole colony. The individual ant and the colony form two halves of a meaningful relationship. There is an interrelationship.

We have begun to see this flavor of unity and interdependent behavior even in the physical world in aspects of quantum theory like non-locality and the interactive relationship between an entity and its environment. We return to Bohm's thoughts:

> The three key features of the quantum theory given do, however, clearly show the inadequacy of mechanistic notions. Thus, if all actions are in the form of discrete quanta, the interactions between different entities (e.g., electrons) constitute a single structure of indivisible links, so that the entire universe has to be thought of as an unbroken whole. In this whole, each element that we can abstract in thought shows basic properties (wave or particle, etc.) that depend on its overall environment, in a way that is much more reminiscent of how the organs constituting living beings are related, than it is of how parts of a machine interact. Further, the non-local, non-causal nature of the relationships of elements distant from each other evidently violates the requirements of separateness and independence of fundamental constituents that is basic to any mechanistic approach.[6]

Additionally, it should briefly be mentioned that another glaring issue with our current model is that it is not even counting all the parts! Again, we are only counting *material* parts, those pieces that are physically visible and can be measured. But if we are talking about all the individual pieces that go on to construct *reality*, we've left a lot out! If we are considering a total system to explain everything and looking at all the elements that compose it (which is the essence of any ontology), we again come up against the pesky reality of the subjective observer.

> ...In a more detailed description the atom is, in many ways, seen to behave as much like a wave as a particle. It can perhaps best be regarded as a poorly defined cloud, dependent for its particular form on the whole environment, including the observing instrument. Thus, one can no longer maintain the division between the observer and the observed (which is implicit in the atomistic view that regards each of these as separate aggregates of atoms). Rather, both observer and observed are merging and interpenetrating aspects of one whole reality, which is indivisible and unanalysable.[7]

If the observer affects the material environment in any way, whether through the observer effect we've discovered in quantum mechanics, the synaptic plasticity we see in neuroscience, or the epigenetic expression we see in biology, or any other way, then we must consider it within the overall sum of interactions that take place in reality. From our ontological outlook, not only are the parts essentially separated from each other (and the whole not considered in and of itself), but the observer's role within the whole sum of interactions is not counted either.

The causal order matters, because elements that occur later in a sequence are necessarily dependent on the ones that come before them. If the observer is affecting change in the physical world, there is a causality there that must be understood within the greater context of the overall sum of interactions. If the organic comes before the mechanical, then elements of the order of living systems like interdependence and structural unity must be considered within the greater structure of the cosmos. If the whole precedes the parts, then there is a unity behind all dual relationships which comes with its own set of qualities, apart from those possessed by each of its poles.

FROM UNITY TO DUALITY

The world of divided objects descends from wholeness. Division, which is the creator of parts, always starts from a unity. In the same way that you cannot have a part without a whole to which it belongs, *all dualities descend from unities*. To be able to see this underlying unity is essential to understanding reality. The implication is that any polar relationship between two naturally occurring opposites is not actually speaking of two separate entities, but one entity expressed in two ways, two faces of the same structure.

Let's take temperature as an example. We tend to imagine a duality to temperature between hot and cold. Hot and cold seem to be opposites. But in reality, hot and cold don't exist objectively. They are relative terms that depend on our position on a scale, our frame of reference. What is "hot" for one observer may be "cold" for another whose position is further away from zero. Hot and cold are simply words that describe relative degrees of difference on the scale of temperature. They are not fixed or absolute terms that describe objective features of nature, but relative terms that are defined relative to our position on a scale. The unity behind these two is the scale of temperature, one structure to which they both belong, representing opposite poles of the continuum.

The mistake we often make in our conception of reality, whether within personal experience or our conception of the cosmos, is in thinking of the parts as separate. It is true that parts always have an element of individuality. They are autonomous and whole unto themselves (a sub-totality, if you want), and this is very fine. *But to think of them as absolutely separate is to make a logical error.* Parts cannot be absolutely separated from a whole.

It is for this reason that the entirety of the cosmos must first be thought of as a unity. (Even if it can also be considered in its parts.) In our current ontological models, we often start with parts and try to *create* a unity. We try to *synthesize* reality from its parts. But the truth is, you cannot start reality from a fragment.

When it comes down to it, the whole is more than the sum of its parts. Parts are parts. They have their own purpose and function, and in a way, their own identity. But the same is also true of the whole. You cannot simply "add up" all the parts and pretend to have a complete understanding of a thing. The whole has its own qualities. It must be considered in its own right as a separate entity. Parts give us insight into mechanics, order, structure. They pertain to the intellectual, quantitative, and compositional. They speak to division, duality, and multiplicity. The whole gives us insight into nature, purpose, and identity. It pertains to the qualitative and the sensory. It speaks to unity. These two exist in an interrelationship. They are two sides of the same coin.

In nature, there is often a pattern of things beginning in simplicity, moving into complexity, and then returning to simplicity again. This is visible in many things, such as the human life cycle (childhood, maturity, and old age). *I believe the same will be true of our science.* We began in simplicity and have now spent an inordinate amount of time invested in the details, trying to understand the varying complexities of nature, specializing and analyzing her "parts." We have had time to fully investigate certain branches of knowledge in extensive detail. But it will eventually be time to zoom back out again and look at the whole picture. It will be time, in observing its properties revealed in wholeness that cannot be seen in its parts, to understand the meaning and purpose of the whole display. It will be a time for us to return to the fundamentals of natural law, the foundation upon which all of reality is based.

CHAPTER 14

THE CONTINUOUS AND THE DISCRETE

The duality between the whole and the parts is related to another duality, visible in physics, that is essential to understanding reality. *The continuous and the discrete.* Canadian philosopher and mathematician John Bell does a good job of laying this out:

> We are all familiar with the idea of continuity. To be continuous is to constitute an unbroken or uninterrupted whole, like the ocean or the sky. A continuous entity—a continuum—has no "gaps". Opposed to continuity is discreteness: to be discrete is to be separated, like the scattered pebbles on a beach or the leaves on a tree. Continuity connotes unity; discreteness, plurality.[1]

In many ways, discreteness is essential to the viewpoint that our scientific model puts forward. Atomic physics takes a viewpoint of the cosmos as composed of fundamental particles that come together to form the larger, conglomerate objects we witness in nature. The idea is that by understanding these fundamental pieces or particles, we can then understand the whole.

Discreteness, which implies clear countable parts, is perfectly suited to the quantitative, mathematical treatment of nature that is dominant in our sciences. It is also well suited to the use of reason and the intellect, which like to divide and order. Continuous phenomena, on the other hand, are not countable in the same way. They are not as clearly defined with regard to their boundaries and therefore require different mathematics to deal with that are often less precise.

Mathematics, being the principal tool of our quantitatively-based, object-oriented science, is ideally suited to the discrete. Mathematics is dependent on clear, countable units to do its work. While it can work with more "cloud-like" objects that have less clearly defined boundaries, such as the "electron cloud" for which we use the wave function in quantum mechanics, it loses precision. This type of math, rather than being exact falls more into the realm of probabilities and potentials rather than certainties. Furthermore, in these situations we must focus on discrete aspects of the system (such as energy levels, spin states, orbitals, etc.) in order to render the situation analyzable with quantitative tools. Qualities are reduced to quantities such that everything can be dealt with discretely. Bell continues:

> The realm of the continuous is traditionally associated with intuition, that of the discrete, with reason. The discrete is a model of tidiness in which quality is reduced to quantity and over which the concept of number reigns supreme. Populated by units lacking intrinsic qualities and so wholly indistinguishable from one another, in the dominion of the discrete difference is manifested through plurality alone. The simplicity of the principles governing discreteness has recommended it as a paragon of intelligibility, a realm within which reason can be realized to its fullest extent. By contrast, the continuous is a jungle, a labyrinth. It teems with such exotic and intractable entities as incommensurable lines, horn angles, space curves, one-sided surfaces. The taming of this jungle by reduction to the discrete has been a principal task, if not the principal task, of mathematics.[2]

The rational mind is attuned to order. The intuitive mind to flow. These two are complementary aspects of reality which are visible in the qualitative difference between the continuous and the discrete. To illustrate the nature of this duality, it is helpful to turn to geometry.

THE DISCRETE AND THE CONTINUOUS
IN GEOMETRY AND COSMOLOGY

There are many ways we can demonstrate the duality of the continuous and the discrete in mathematics, but one of the most simple is the real number line. The real number line is a line that extends infinitely in two directions. It has its center (or origin) in zero, with negative values on one side and positive values on the other. If you move from zero in any direction and stop at any point along this line, you have arrived at a discrete entity. A clear, countable value. Sometimes it is a whole value (like 1, 7, or 143, for example), and sometimes it is a fraction (like 1.5 or -13.2).

But if you consider the distance between this value and zero, you cannot really say that there are a discrete number of units between them. The line itself is a continuous entity that can potentially be divided an infinite number of times. Between the number zero and one for example, there are potentially an infinite number of discrete entities. All you have to do is keep cutting your value in half, for example, and you can see this. The discrete entities, the countable numbers, are actually descendants from the continuous entity, the whole line.

Another way to think about this is to consider the difference between the two most basic geometrical figures, the point and the line. The simplest geometrical figure is the point, which we call zero-dimensional. The fundamental particles in the Standard Model of particle physics are idealized versions of these called *point particles*. What this means is that, at least within the idealized forms that science uses in its mathematics, *they have no extension in space*. They are essentially a coordinate, but do not really "take up space." When we increase in dimensionality from the zero-dimensional point to the one-dimensional line, certain irregularities arise.

There are two primary ways we can view one dimension. *These two ways of looking actually reveal two fundamental ways in which we must view the nature of reality itself.* In the first way, we imagine that the lower dimensional object, in this case the point, is repeated infinitely along the higher dimension's axis. In the case of the point, we imagine it is repeated over and over again in two directions infinitely, with no gaps in between them. In this picture of things, *the line is a composite of points.*

This process of repeating a lower-dimensional object along a higher-dimensional axis can be repeated for any increase in dimensionality. For example, from one to two dimensions, we can repeat the infinite line

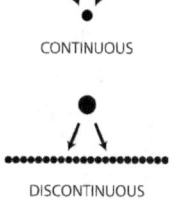

CONTINUOUS

DISCONTINUOUS

over and over again in two new directions to get a plane. From two to three dimensions, we can endlessly repeat a plane to get 3D space. What is important to note is that in all of these cases, all of the objects (the lines, planes, and 3D objects) are ultimately reducible to these same zero-dimensional particles or points. From this point of view, *the parts are primary*, because they are the basic constituents from which all other things are made. You could say that in this viewpoint, the cosmos is considered to be *discontinuous*.

In the second viewpoint, rather than the line being composed of separate individual units, it is one continuous object. There are no gaps or separations between any of its "parts." Everything exists as *one* structure. This structure can then be *divided* into pieces and parts after the fact, in order to examine some segment of the continuity more closely. In this viewpoint, rather than the cosmos being a conglomeration of individual points that *look* like one object, it actually *is* one object, with each individual point being an abstraction from an undivided whole. In this viewpoint, the cosmos is essentially *continuous*.

FIELDS AND QUANTUM FIELD THEORY

The conception of the universe as discontinuous has serious logical inconsistencies. To begin with, if a point is truly without extension, then no matter how many times you repeated it, it could not have "moved" anywhere! The word "extension" means to extend, stretch, or reach in some direction. If the point is without extension, that means it cannot do this. If it *does* have some extension, then it is no longer zero-dimensional and would be considered at minimum a line! Therefore the concept of a line made of an infinity of points does not really make sense, especially since it requires you to consider its composition by moving from the lower dimension to the higher one. Yet moving from the higher dimension to the lower one does not lead to these same inconsistencies. You can abstract a point from any position on a line without creating any problems.

Until fairly recently, ideas of discontinuity and particle-based theories of the universe dominated our physical cosmology, but that changed with the introduction of *fields*. Fields are exemplary of the process of moving from continuity to discreteness. Unlike particles, fields are not localized to any specific point in space. Instead, *they exist everywhere at once*. Fields have a numerical value not just for one point, but for *every* point in space-time. In this way you can consider the particle and the field to be diametrically opposed to each other. This is with respect to the spectrum between a fully continuous entity that is totally dispersed throughout space, and a fully localized entity with a concrete, specific value.

The duality between the continuous nature of fields and the discrete nature of particles is the clearest physical example we have of this basic cosmological principle. They represent essentially two types of primary entities in physics: discrete entities with specific, localized positions in space-time which we idealize as material points (and which form the elementary particles of the Standard Model of particle physics), and continuous phenomena that don't have specific locations but are instead dispersed across wider areas of space like a field or a wave.

In quantum field theory (QFT), one of the most prominent theories of quantum mechanics, each elementary particle, rather than being a free-standing entity disconnected from everything else around it, is actually a concentration or "excitation" of energy somewhere within the field. This means that the particle is not a free-standing entity floating around in empty space, but is an *emergence* from a greater continuous entity, like a wave emerging from a body of water. In QFT, all of the fundamental particles and forces of physics each have a field to represent them. German-American physicist Henry Margenau comments, bringing in some of Einstein's own thoughts:

> Pre-quantum physics was marred by a peculiar dualism of conception, the irreconcilability of particles with fields, or, in more fundamental terms, the contrast between the discrete and the continuous. The particle notion received its confirmation at the hands of Newton and culminated in the brilliant speculations of Helmholtz. But the very idea of a particle becomes logically unsound unless it is stabilized by an absolute space or an ether to provide an invariable reference for its instantaneous position. The situation is well described by Einstein whom we quote at length.
>
> > —Before Clerk Maxwell people conceived of physical reality—in so far as it is supposed to represent events in nature—as material points, whose changes consist exclusively of motions, which are subject to ordinary differential equations. After Maxwell they conceived physical reality as represented by continuous fields, not mechanically explicable, which are subject to partial differential equations. This change in the conception of reality is the most profound and fruitful one that has come to

physics since Newton; but it has at the same time to be admitted that the programme has by no means been completely carried out yet. The successful systems of physics which have been evolved since rather represent compromises between these two schemes, which for that very reason bear a provisional, logically incomplete character, although they may have achieved great advances in certain particulars.

The first of these that calls for mention is Lorentz's theory of electrons, in which the field and the electrical corpuscles [particles] appear side by side as elements of equal value for the comprehension of reality. Next come the special and general theories of relativity, which, though based entirely on ideas connected with the field-theory, have so far been unable to avoid the independent introduction of material points and ordinary differential equations.

The author's preference is here very clearly stated. Reality is to be regarded as a continuous manifold.[3]

According to QFT, matter itself is actually a quantum field (despite a field not being a physically visible entity). Though it is "invisible," it has energy, exists in space, and has properties that can be calculated and accurately predicted by experimental results. All of the elementary particles (the smallest particles physics has been able to identify so far) *are* quantum fields, with the "particle" part being excitations or concentrations of energy within it. Every particle in the universe is a ripple, excitation, or bundle of energy of the underlying quantum field. Photons, for example, are particle equivalents of the electromagnetic field, electrons of the electron field, and so on. This means that physical reality has a continuous nature within which we can witness discrete entities, but the discrete entities are not primary in the causal order. Even space-time we call a continuum.

The introduction of fields revolutionized our view of the real world. It firmly established the idea that physical reality was one continuous manifold, an undivided whole that could then manifest itself into smaller parts by way of excitations or activities within some part of its continuous structure. The idea of free-standing particles floating around in nothingness, an idea first born in ancient Greece many thousands of years prior, was finally being replaced.

THE CAUSAL ORDER - UNITY TO PLURALITY

Continuity is primary to discreteness in the causal order for much the same reason that the whole is primary to the parts. You cannot start from a fragment because a fragment implies a greater whole to which it is part. In the same way, any "discrete" entity must exist within a greater fabric from which it is abstracted or generated, like a particle from a field. The implication of continuity being primary in the cosmic order is that at the root of our universe is an undivided wholeness. Bohm comments:

> Einstein did in fact very seriously try to obtain …a description in terms of a unified field theory. He took the total field of the whole universe as a primary description. This field is continuous and in-divisible. Particles are then to be regarded as certain kinds of ab-straction from the total field, corresponding to regions of very in-tense field (called singularities). As the distance from the singularity increases, the field gets weaker, until it merges imperceptibly with the fields of other singularities. But nowhere is there a break or a division. Thus, the classical idea of the separability of the world into distinct but interacting parts is no longer valid or relevant. Rath-er, we have to regard the universe as *an undivided and unbroken whole.* Division into particles, or into particles and fields, is only a crude abstraction and approximation. Thus, we come to an order that is radically different from that of Galileo and Newton—the order of *undivided wholeness.*[4]

A movement from a whole (continuous) to parts (discontinuous) is also a movement from unity (one) to plurality (many). To take a whole and split it up into smaller pieces creates multiplicity from a whole. The relationship between the whole and the parts, and between the contin-uous and the discrete, is also at the same time the relationship between "the one" and "the many." Between unity on one hand and multiplicity on the other. Everything begins in unity and then descends into parts. It is the continuous, undivided whole that is first and illustrates unity. Then, when we focus on any individual area of its continuous structure, we have a part. Plurality is generated from unity.

With that said, we do not have to think of the continuous and the dis-crete as mutually exclusive. They are not irreconcilable. Just because they are opposites in nature, that doesn't mean that they can't or don't appear together. We have already seen this in one of the revolutionary ideas of the early 1900's, wave-particle duality. Elementary particles likė photons and electrons, while being abstractions from the field, can also manifest themselves as either a wave or a particle. They do not simply have one nature that never changes. While we debated for millennia whether light was particulate and discrete or wave-like and continuous, we finally settled the debate in the early 1900s. Light was neither a wave nor a particle. It was both.

Since then we have discovered that this is true of *all* matter. Theoretically speaking, all matter can behave as either a wave or a particle which we have dubbed "matter waves." This illustrates the fact that the discrete and the continuous are not binary in nature, but instead exist on a spectrum. Observing the properties of one brings those qualities out, and vice versa. It exemplifies a principle that is hugely important in our analysis of the real world called *complementarity*.

In truth, *all discrete entities are both discrete (particular) and continuous (wave-like) at the same time.* That is to say, they descend from the whole, the universal field, but are also possessed of their own individual nature. This is an observable constant in the universe. Every "thing" we call individual, that has some inward integrity and can be called a "unity," has at the same time a discrete quality, that which makes it individual and unique, and a continuous quality, that which connects it to the greater whole from which it descends. What is important to realize is that this is a fundamental of natural law. All dual or polar relationships, rather than being mutually exclusive, are *complementary*.

CHAPTER 15

COMPLEMENTARITY

The principle of complementarity was first brought to light in the sciences by Niels Bohr in the early twentieth century. On a skiing trip to Norway, he received a letter from Werner Heisenberg outlining his as-yet-unpublished theory about uncertainty in physical systems, what would later be known as the Heisenberg Uncertainty Principle. This prompted Bohr to consider that this uncertainty may be the result of a core relationship between opposite but complementary forces in nature.

One of the ways this complementarity revealed itself was in the dual wave-particle nature of light. The more wave-like qualities were observed, the less visible particle-like qualities became, and vice versa. Continuity and discreteness were another visible dualistic, polar relationship that lent themselves well to the idea of complementarity, the former being exemplified by field theory and the latter by atomic physics. Nevertheless, Bohr did not stop at physics. He identified complementarity in many other fields, including biology (in the duality between the organism and its environment) and most notably, between the subject and the object.

DUALITIES AND POLAR RELATIONSHIPS

In our broad and far-reaching exploration of the nature of reality, we have identified many dualities. Let us revisit some of them. One of the most important of these is between the subject and the object, or, stated in other terms, between the observer and their material environment. This we can see from two perspectives. Either we can see it from the subjective perspective (from the perspective of our own unique and individual frame of reference), or we can look at it objectively (from the perspective of a "disembodied" intellect).

From a subjective viewpoint, this relationship is very obvious. We have our own inner life and all the various nuances that go along with it. Our feelings and emotions, our mind and thoughts, our physical bodies, the relationships we have with friends, family, and partners. This "inner world" is clear and visceral to us as subjective beings. But it is also obvious that we live in a physical matrix that comes with its own set of laws. When we drop an apple from our hand, it falls to the ground. When our vehicle accelerates, it is harder to slow it down. The material environment that surrounds us has rules, and this is visible to us. So from the subjective perspective looking outward, we can see a relationship between "the subject" (us), and "the object" (the environment).

From an objective viewpoint, we can also view the relationship between the subject and the object. For example, we can see that the subject is ubiquitous, everpresent. No matter how or when we analyze, the "analyzer" must necessarily also be there. Whenever we make an observation, there is always an observer. This observer always has a particular position in space and time. A frame of reference if you want. This is an objective truth. It is true in all instances and at all times (except when we idealize or imagine otherwise in our minds). It is also clear that this "observer," with their particular frame of reference, exists within a physical matrix that has rules. By sticking to a strictly objective viewpoint in which we reduce or eliminate the observer and the qualitative elements of our reality, we can render the whole cosmos subject to reason, quantification, and the intellect's ordered mind. Essentially, we can do math with everything.

It is clear from these observations that the subject and the object are ubiquitous in nature. The only place where one or the other disappears, is in our imagination. In fact, this is one of the roles of the intellect and objectivity. It provides a "free place to play." We can imagine anything we want, including a reality without ourselves in it. Objectivity comes from the mind. This objectivity, the impersonal and indirect, perfectly complements our subjectivity, the personal and direct. Those aspects of existence that allow us, from within our own frame of reference, to interact with the matrix. These two sides exist in an interrelationship. They are complements.

This interrelationship between subjectivity and objectivity, between the subject and the object in nature, is also mirrored by the duality between sensory experience and intellectual analysis. Sensory experience is a direct and personal way of gathering information about the world. When we touch a sweater we want to buy at the mall, we are learning about this object in a direct, personal way. The intellect can analyze things in a way that is impersonal and indirect, from an "objective" place that is outside of the frame of reference of the individual observer.

The sensory and the intellectual, the bodily and the mental, take us into yet another duality between feeling and thinking. While thinking is strongly correlated to the mind and mental faculties, feeling is strongly correlated to the body and sensory faculties. Feelings and thoughts form a central duality in the observer's daily experience, but are also important in the science we do and the model that arises from it. In our current model, things are based on thought and analysis, but feeling and sense are largely absent.

The feeling and the sensory, being more based in the body than the mind, take us into yet another duality, that between the organism, the living thing, and the machine, pure mechanism. This could also be phrased as the duality between the animate (living) and the inanimate (non-living). In many ways, physics and biology are scientific representatives of this duality, as physics deals with the mechanics of "dead" objects, whereas biology deals with the mechanics of living organisms. We have discovered that living organisms possess qualities that cannot be cleanly quantified like a "will-to-live," purpose, and individuality. Like with objectivity, principles and qualities in the inorganic world can be universally applied and speak in generalities, whereas the world of the living organism always possesses an element of individuality that makes it personal and sets it apart from inert objects.

Organisms also exhibit a sense of wholeness and inward integrity that is reflected in the integrative behavior of their cells, each of which possesses its own identity while at the same time operating within the context of a greater whole. This behavior of being "individual" and "collective" demonstrates how organisms exhibit "oneness" and "manyness" at the same time, each principle coexisting with and supporting the other. This principle of wholeness visible in organisms speaks to a broader phenomenon within the cosmos, in which there is always a relationship between whole entities and the individual elements that go on to compose them.

While the parts each have their own role and function within the whole, the whole itself is also possessed of qualities that cannot be discerned by looking at the parts alone. Qualities like beauty or individual identity, for example, are visible in wholeness, but not when looking solely at the parts. In order to understand something in its entirety then, both sides of this equation must be understood and incorporated at the same time.

The principle of wholeness and its primary occurrence within the causal order illustrates yet another duality that became visible with the advent of quantum field theory in the 20th century. The duality between continuous entities which have no "gaps" between them but exist in an unbroken whole, and discrete entities which are concentrated into points like grains of sand. The duality between continuity and discreteness is reflected even in physical systems, like that between fields or waves and particles. These dualities imply yet another, that between unity and plurality, what is sometimes referred to as "the one" and "the many."

In the causal order and the unfolding of the cosmos, we see the precedence of the whole before the parts and the continuous before the discrete. We see that "manyness" proceeds from "oneness," plurality from unity. When we connect this with our understanding of organic and inorganic systems, we begin to see that the physical elements of the cosmos (especially as revealed through quantum theory) bear more resemblance to the way organs in a body interact than the way lifeless entities that interact externally and bear no "internal" connection to each other do. This seems to support a vitalistic rather than mechanistic conception of nature, in which nature as a whole is considered a living, breathing entity.

THE UNITY OF OPPOSITES

Because dualities are literal opposites, it is easy to think of them as having nothing to do with each other. East is the opposite of West, hot is the opposite of cold, etc. On the surface, it seems as if they couldn't be more different from each other. Yet this is only from a surface-level perspective. As we previously established, consider that the words "hot" and "cold" are actually relative statements that depend on your frame of reference within the scale of temperature. What is "cold" within your frame of reference may actually be "hot" to another observer whose frame of reference is closer to zero. In truth, there is no such thing as "absolutely hot." Hot and cold describe a relative state, depending on your frame of reference. In truth, *they both represent degrees of difference on the unified scale of temperature.*

Similarly, east and west are relative terms. Despite the fact that they seem to be referring to opposites, they are both referring to a scale of directionality. If I go west far enough, I eventually arrive east of where I started. All dualities are like this. The polar relationships are simply describing degrees of difference within the same scale of measurement. Therefore, despite the poles of a duality seeming to be opposite each other at the surface level of things, at their root, they are part of the same underlying structure. Like the two faces of a penny.

The implication here is that polar opposites are not mutually exclusive (meaning that they can't exist together), but *complementary* (meaning that they complement and complete each other). Hot makes cold possible, east only exists because of west, light can only be known because of darkness, etcetera.

What seem like opposites are not absolute and isolated entities, but are rather representatives of *the polar extremes of a continuum*. We as the "player" in the game, so to speak, the subject within the field of objects, are always in a relative position within this continuum, and the words we use to represent the poles help us to isolate our relative position within it. In this way, all polar terms (hot and cold, soft and hard, fast and slow, etc.) are relative to our position within the spectrum they represent. This continuum is the structural unity beneath what seem, on the surface, to be mutually exclusive opposites.

THE RECONCILIATION OF PARADOX

During the time of the Enlightenment we made the decision, albeit unconsciously, to exclude subjective elements of reality from our study of the whole cosmic order. Using the scientific method and a strict adherence to objective viewpoints, we believed we would come to answers. But this was never meant to provide answers to *everything*. This method has been successful in discerning natural laws as they pertain to physical reality, but when it comes to the other half of reality pertaining to the subjective observer, we have remained largely in the dark.

The subjective observer is a complementary opposite to the reality of the observed object within the whole reality of experience. Subject and object are two halves of a meaningful whole. The idea that one can exist without the other has major logical inconsistencies. The subjective and the objective, the sensory and the intellectual, feeling and thought, the organic and the mechanical, the living and the non-living are each complementary opposites that speak to a grander design within the cosmos. A duality between an experiencer, a "subject" within the game so to speak, and all the rules, structure, and limits that make their experience possible.

The whole spectacle of created reality is only possible because of duality. If everything was unity, there could be no dynamics through which reality could be experienced. By realizing the complementary nature of all dualistic structures, we begin to reconcile the many paradoxes that seem to plague our ontological models. The reconciliation of paradox comes from straddling the line between the intuitive and the rational. Between the personal and the impersonal, the subjective and objective aspects of our observable reality. The animate and the inanimate are an integral whole. The more you go to one pole or the other, the more you draw that reality out. Clarity can be found by acknowledging their validity simultaneously, while at the same time perceiving the underlying unity between them.

The answer to our deepest questions about the universe lies in the reconciliation of paradox. To hold two simultaneously existing opposites in the same breath, and to perceive their connection. Opposites only really make sense in the context of each other. They are born together from one unity. This is just as true for physical things as it is for metaphysical ones. In fact, it is the physical and the metaphysical, the formless and the formed, the material and the immaterial, the measurable and the immeasurable, that are the greatest duality of all.

CHAPTER 16

THE MEASURABLE AND THE IMMEASURABLE

The idea of two complementary opposites which come together to form the world is not new. One of the most famous examples of this is the yin and yang of Chinese Taoism. In Taoism, there are two fundamental forces that mutually govern the universe. Yin represents the feminine element, the force of receptivity and inwardness. She also represents the black, that which is formless and immeasurable. Yang represents the masculine element, the force of outwardness and expression. He also represents the white, that which is manifest and measurable. These two forces exist by each other's grace. They create and complement each other.

But yin and yang are simply words meant to point out a deeper underlying reality, something that goes beyond thought systems and ideologies. They are meant to point to a deeper science of energy beyond mere forms and material objects.

THE EAST AND THE WEST

A simple way of understanding the relationship between these two fundamental forces is to consider the difference between the measurable and the immeasurable. That which is measurable is fairly easy to understand.

It is the basis of all our science, and consequently, our basic outlook on the real world that has arisen from it. The process of measuring and weighing is especially characteristic of the Western approach to reality. To illustrate this further, we turn to the thoughts of physicist David Bohm:

> ...It is useful to begin by going into the difference between Western and Eastern notions of measure. Now, in the West the notion of measure has, from very early times, played a key role in determining the general self-world view and the way of life implicit in such a view. Thus among the Ancient Greeks, from whom we derive a large part of our fundamental notions (by way of the Romans), to keep everything in its right measure was regarded as one of the essentials of a good life (e.g. Greek tragedies generally portrayed man's suffering as a consequence of his going beyond the proper measure of things). In this regard, measure was not looked on in its modern sense as being primarily some sort of comparison of an object with an external standard or unit.
>
> Rather, this latter procedure was regarded as a kind of outward display or appearance of a deeper 'inner measure', which played an essential role in everything. When something went beyond its proper measure, this meant not merely that it was not conforming to some external standard of what was right but, much more, that it was inwardly out of harmony, so that it was bound to lose its integrity and break up into fragments.[1]

Therefore, whether referring to the "inward" characteristics of lived experience or the "outward" characteristics of physical objects, proper measurement was seen to be one of the primary tools in creating and maintaining harmony. Because of this emphasis in the West, measurement became paramount in all things, including our science.

At its essence, the concept of measure has to do with *limits*. As previously established, all things begin in continuity and wholeness. From this continuous entity, like a line for example, we can subdivide it into smaller ordered segments. We can create some sort of regular interval that we define in some way. To be able to "measure" then, has to do with the choice we make about how to subdivide a larger continuous entity into smaller, discrete pieces or parts.

> ...Measure has to be specified through proportion or ratio... thus, to set up a scale (e.g. of length) one must establish divisions which are

in effect limits or boundaries of ordered segments.[2]

These "intervals" must be based on something. There are truly no "objective" measures in nature. Everything is in motion, constantly changing. Therefore in order to get a "standard of measure," we must base that ordered segment on something that doesn't change very often. We base a "meter" for example on the length of the path traveled by light under very specific circumstances (in a vacuum, during a time interval of 1/299,792,458th of a second). We base measurements of time on the rotation of the Earth around the sun, and so on. Units of measure are ultimately at the discretion of the observer, based on the way we choose to divide up a continuous entity. *All standards of measure are ultimately relative.* We try to base our units on elements of the physical environment which are relatively stable, but how we choose to carve up the continuous entity into smaller, discrete pieces is ultimately at the discretion of the subject.

All of this is possible because of the observer's ability to set limits. Within the philosophy of the Tao, yin pertains to the limitless, and yang to limits. Nevertheless, the centrality of measure, limit-setting, and the rational approach to reality was particularly emphasized in the West. In the East, this was not the case. In the East, notions of limitlessness and the immeasurable side of this fundamental dualistic relationship were primary in the causal order of nature. Bohm continues:

> Now, in the East the notion of measure has not played nearly so fundamental a role. Rather, in the prevailing philosophy in the Orient, the immeasurable (i.e. that which cannot be named, described, or understood through any form of reason) is regarded as the primary reality. Thus, in Sanskrit (which has an origin common to the Indo-European language group) there is a word 'matra' meaning 'measure', in the musical sense, which is evidently close to the Greek 'metron'. But then there is another word 'maya' obtained from the same root, which means 'illusion'. This is an extraordinarily significant point. Whereas to Western society, as it derives from the Greeks, measure, with all that this word implies, is the very essence of reality, or at least the key to this essence, in the East measure has now come to be regarded commonly as being in some way false and deceitful. In this view the entire structure and order of forms, proportions, and 'ratios' that present themselves to ordinary perception and reason are regarded as a sort of veil, covering the true reality, which cannot be perceived by the sense and of which nothing can

be said or thought.

It is clear that the different ways the two societies have developed fit in with their different attitudes to measure. Thus, in the West, society has mainly emphasized the development of science and technology (dependent on measure) while in the East, the main emphasis has gone to religion and philosophy (which are directed ultimately toward the immeasurable).[3]

The key difference between these two perspectives is in what they consider to be the essence of reality. On the one hand there was the belief that the essence of reality was in the measurable, that which was manifest and could be seen. On the other hand was the belief that the essence of reality was in the immeasurable, that which could not be seen but was tangible through experience.

THE IMMEASURABLE ASPECTS OF REALITY

To make this more tangible, let's identify some of the things within our reality that can be identified as belonging to the immeasurable side of this duality. Most of the things that could be classified as immeasurable are those that can be cognized through conscious perception, things that might be dubbed subjective but which are not strictly physical. We have already identified several of these.

Qualitative aspects of reality for example, experiences we have through the senses like color, sound, taste, or smell, all qualify. The taste of bitterness cannot be quantified, but is a real phenomenon. The qualitative experience of color like cobalt blue cannot be quantified or measured with physical instruments.

The phenomenon of beauty (or more generally "aesthetics") belongs totally to the experiencing subject. While we know that it is a real phenomenon and essential to the experience of life we have as observers of nature, it is totally outside of the scope of physical measurement or quantification. We cannot do math with beauty.

Another important one we have already discussed is volition or the will. The will as a phenomenon is not a physical object that can be measured and made subject to quantitative or mathematical analysis. The will always belongs to a subject, to that half of the overall sum of reality. The will is immeasurable but tangible, an important component in the overall sum of interactions that go on to compose the world.

Another we have covered so far, so important in the overall sum of interactions that make our living world what it is, is memory. While we have established that, to some degree, memory is recorded in the physical organism, and even to some degree in the DNA to preserve some semblance of continuity in the evolving lifeform, the *experience* of memory for the living subject is something else entirely. Even the phenomenon of memory itself could not be possible without a subject to do the remembering, because it is based entirely on experience. Memory is the impetus for change, but is itself an immeasurable phenomenon, existing outside of the world of strict physicality and belonging more to the domain of consciousness and subjectivity.

One of the most important of the phenomena outside of the reach of "the measurable" and most central to our experience of the world is *feeling and emotion*. Feelings and emotions are among the most important and influential phenomena we have, influencing nearly everything that occurs within the universe of subjectivity and relationship. They are the realm of love and fear, happiness and sadness, anger and peace. Everything that makes life worth living is in our feelings, but feelings cannot be seen with the physical eye nor measured with scientific instruments (at least not yet). Feelings are immeasurable. They belong to the living subject.

The phenomenon of value is absent from the worldview of pure physical forms we study in physics and material science. Value, while deeply influential in the everyday lives of everyone and everything that draws breath is also not physically measurable. Value is subjective. It is something we as living beings assign to various things and depends greatly on our way of thinking and perceiving.

What of the phenomenon of care? Are we saying that to care is unreal or that it is simply a byproduct of chemical reactions? Does that mean that we don't really love our children? That our mother didn't really care about us? That all of this is preordained by physics and the chemical interactions between molecules? Even if we did believe that caring was simply a byproduct of chemical reactions, (which doesn't make much sense within the causal order as we have already established), it still wouldn't explain the *experience* of care, which is something else.

Another fairly obvious immeasurable phenomenon belonging to the experiencing subject is imagination. I can imagine a blue bird in my mind, and while another cannot see it, that doesn't mean the experience of the bird is not real for me, personally and subjectively. I can hear Beethoven's Symphony Number 9 in my mind, despite the fact that there are no measurable vibrations in my physical environment to reenact it.

Imagination is a tool we use to create. To simply pretend it is not real or has no value within the overall order of reality is extremely shortsighted.

If something is observable, it is observable. We can idealize reality to bend and distort it into a version that is more convenient for our tools. But reality is there to observe, whether what it presents to us is convenient for our tools and methods or not. Something needn't be physically measurable to exist. We see this in our daily experience. We know it intuitively. From the inside looking out, it is incredibly obvious. But it doesn't fit with our ontological model, the picture of reality that has been created by modern science. A system that has evolved over hundreds of years of looking at reality in a particular way and with a particular set of tools. We have developed a confirmation bias, and it is getting in our way. A bias brought on by a version of science, 500 years strong, which has become so enamored with its own success that it has grown rigid and inflexible.

It is precisely these immeasurable, conscious aspects of our reality that are most in chaos in our observable reality. Does it seem like an accident in the context of the greater dualities that we have discussed that the side we have ignored is the one that has manifested the most chaos? What is this chaos if not a reflection of our lack of understanding when it comes to the order of reality pertaining to subjectivity, our inner life, and relationships?

If the material world is one half of the overall reality of energy, then material science is only one pole within a greater science of energy.

The immaterial and immeasurable, the reality of the experiencing subject and the laws that govern their experience are the other half of this science. These laws pertaining to the subject and their interaction with the material world (and other subjects) are what is missing from the total order of reality which must be accounted for in our model.

The truth is, nature without an "experiencer," a subject, doesn't make a lot of sense. It is pointless. Meaning is something starkly absent from our entire assessment of reality. Meaning and purpose are also immeasurable elements of the cosmos. The meaning and purpose of a thing may not be physically measurable, but at the end of the day, they are among the most important aspects of our observable reality. Meaning and purpose point to the question of *why*.

Science, as we practice it today, doesn't really care about why. It is concerned with what, where, when, and how, but *not* why (or who). This it dismisses as "philosophy" and renders irrelevant to an overall assessment of the cosmic order. Does that not seem problematic? Meaning, the *reason* behind why things happen, reveals their purpose, their end. Does this seem irrelevant in the grander scheme of things? An ontological system, a metaphysical outlook that does not consider the living subject (the who) and excludes it as a rule, *will never really understand purpose* (the why).

In French there is a phrase—"raison d'etre"—which translates to "reason to be." A basic, essential purpose. A reason to exist. A motivation. This purpose is as essential to life as breathing, but because it is immeasurable, it exists outside of the purview of our science, our *truth*.

Meaning and purpose arise from consciousness, from the immeasurable. They are born from the living subject. The cosmological outlook of the science of today, an incomplete and incoherent science within the grander scheme of things, at least from the perspective of the observer's point of view looking outward, is missing all the *contents* of life. It is concerned with its form and structure. Its physical mechanics. But so many of the things that make reality what it is to the living observer are simply missing. There is no color or beauty. No free-will or choice. No memory or experience. No purpose or meaning. It is flawless, but empty. Functional but soulless. That's a problem. Because if we hope to create harmony for our world, we must consider everything that exists within it. We must come to an understanding which can address our whole experience. This requires from us a willingness to be flexible and to change. But, above all else, it requires a commitment to the *truth*.

PART V

TRUTH

CHAPTER 17

CONSCIOUSNESS

In modern times, scientists have become the new gatekeepers of truth. In older ages, this was primarily the role of priests and religious intermediaries. If someone had a question about the nature of the universe, they would confer with them in temples or holy places. Now the temple has become the laboratory and divination the scientific method. The label "backed by science" is enough for most people in the modern world to give validity to virtually any claim. Still, despite its successes, the nature of consciousness stubbornly resists integration with our models.

Questions of truth used to be broader, encompassing everything from right moral behavior to the nature of the soul to the path to enlightenment. Now these have been relegated to other fields. We often resign ourselves to the idea that we will never really come to answers with these things. As conscious and physical elements of our worldview have fractured, so too has our *truth*. Much of this has to do with consciousness.

Beyond the forms and structures visible in the physical world, there is the whole universe of conscious, inner experience that evades measurement by physical tools. Understanding how to fit this immeasurable, experiential element into the total order of reality has been a problem for science. In fact, this problem was given a name in the mid-1990s by Australian philosopher and cognitive scientist David Chalmers. He called it "the hard problem of consciousness," distinguished from the "easy" problem by the fact that conscious experience cannot be measured in the same way physical things can.

THE HARD PROBLEM OF CONSCIOUSNESS

One of the challenges to understanding consciousness has to do with how we frame it within the total order of reality. It is commonly assumed in academic and scientific circles that consciousness is an emergent property of matter. The most common theory about how this happens is through a property of the community of neurons in the brain and nervous system (see Chapter 11). This is of course based on the assumption that consciousness is produced by the body, and all that is missing is for us to find out how and where this occurs.

It should be clear by now that the idea consciousness is produced by the physical body is an *assumption*, not a fact. We base many of our experiments and reasoning on this assumption, and there is a hint of confirmation bias in this. Confirmation bias is the tendency to favor information that confirms one's preexisting notions. Rather than zooming out and endeavoring to look at the whole order of reality from a more general perspective, we tend to look at it through a physicalist lens. This often causes us to want to *force* consciousness into materialistic models, whether it fits in easily or not.

Here's what we know. There is a material world around us we interact with through our bodies. We experience this material world from within our own subjectivity, our own individual perspective looking outward. This element of subjectivity is always present. The only way we can experience the world without ourselves in it is through an exercise of the imagination. Because this subjectivity is always present, and because we are always experiencing reality from within our own individual perspective, it could be said that this material world is always occurring *within us*.

There is never a time when it is not occurring within us, within our own subjectivity. Through the intellect we can "disembody" ourselves and try to view the world in a neutral way without any of our subjectivity in it, but this always occurs within the mind. It has been a useful tool for certain things, but to call it "the way reality really is," does not follow from logic or experience. Because this "inside-out" situation is valid for all observers at all times, irrespective of any individual circumstances or opinions, we could say that it is objectively true. Reality always occurs within consciousness or the subject. There is never anywhere else it occurs, unless again we imagine it to be so.

Unconsciously, we tend to frame the question of consciousness within a perspective driven by our dominant paradigms and preconceived notions. Because we imagine reality to be physical first, it is difficult to imagine consciousness as being fundamental to reality. From the physical perspective, consciousness seems to be a visitor "from the outside," poking its head into a reality that doesn't really care whether it is there or not. At its root, this is based on the assumption that the object creates the subject. This assumption is so widespread—held by so many people in our collective mind at the same time—that we don't usually see it or think to question it. Our general outlook and ideological philosophy have become defined by it.

ACADEMIC PHILOSOPHIES -
REALISM, IDEALISM, AND DUALISM

In the history of philosophy, there were traditionally three ways of looking at the causal order: realism, idealism, and dualism. In realism, *matter is primary to consciousness*. This is the most widespread viewpoint in the scientific community. Everything is essentially physical first, then all conscious experience and subjectivity emerges from it. In the philosophy of idealism, it is the opposite. *Consciousness is primary.* It comes first in the causal order and all matter, physical forms, and mechanical relationships arise from that. The final viewpoint is called dualism, made popular by Reneé Descartes. It states that reality is fundamentally dual with a reality called "mind" and a reality called "matter" that coexist and depend on each other. In this viewpoint, *neither consciousness nor matter is primary*. Instead, they arise together.

Now before we move on, I want to say a word about some of this terminology. For one, I believe that the words "realism" and "idealism" are stupidly impractical. Language is only as effective as its ability to transmit the right meaning between two parties, and the words realism and idealism have a problematically different meaning in conventional language than they do in philosophy. While this should have prompted the creation of new vocabulary to avoid unnecessary confusion, that was not the case (and often isn't in philosophy, science, and academia which tend to enjoy their specialized vocabulary).

In common language, "realism" is associated with a way of looking at things that is grounded in reality. A realist sees things as they are. There are no flights of fancy, no letting their imagination get away from them.

The word "idealism," on the other hand, is associated with a sort of naive, dreamy way of looking that is disconnected from reality. Someone that does not see things as they are, but as they *should* be. While this is *not* what is meant by these words in philosophy, it is important to point this out. Because while we might understand that their meanings are supposed to be different within the context of philosophy, there is a good chance these words will leave a subtle impression in your mind, potentially biasing you one way or the other because of where the word takes you.

Realism should really just be called "materialism," since its principal tenet is that reality is material first. So let's be clear. We're not talking about "realistic" and "idealistic" ways of seeing the world. We're talking about the incidence of consciousness and matter in the causal order. What are the fundamental constituents of reality, and in what order do they present themselves in its unfolding?

EQUATING REALITY WITH PHYSICS

To reiterate, the majority of the scientific community adheres to the philosophy of realism. The idea that all of reality can ultimately be boiled down to physical components, and that they are the basis upon which all other reality is based. This way of thinking pervades every area of modern society, and is often so deeply rooted in our collective psychology and conception of truth that we do not think to question it (if we are aware of it at all). This materialist view of reality has grown so dominant that often now when we refer to reality, we do not refer to laws of nature, but to laws of physics. *The laws of physics have become synonymous with reality itself.* The implication being of course that reality is fundamentally physical, and therefore all other aspects of reality must be subordinate to it.

We must remember that much of our science is actually based on idealizations, primarily in the realms of physics and mathematics (which serve as a basis for most of the rest of the sciences). These idealizations are not "real" in the strict sense, but are special mental devices that allow us to ignore imperfections in nature to subordinate it to our intellectually-driven models. The goal of course is to make the analysis of certain phenomena cleaner and more manageable for our chosen tools and methods. In reality, these imperfections we get around or eliminate through the process of idealization are actually an integral part of the real world. *They are there for a reason*, and must be understood and accounted for within the context of the overall order of nature.

Ignoring them for the sake of fitting the world into our "perfect" models has consequences, especially for the ontological outlook that emerges from it (see Chapter 10). Much like our physicalist philosophy, the problem has to do with the fact that these idealizations are so commonplace and we are so practiced at using them, that they largely go unnoticed. This is owed in no small part to the predominance of the intellect as the primary tool we use in our analysis of nature. This inclination ensures that the outlook that emerges from its analysis will be biased toward its own way of looking. As we have established, the intellect has particular qualities and characteristics to the way it likes to see things.

We must be careful when leaning too heavily on only one of the tools we have available to perceive things, and to be conscious of their limitations. By using the full spectrum of the tools we have at our disposal to perceive reality, we can achieve a more balanced view of the whole.

THE EVOLUTION OF CONSCIOUSNESS

In understanding the role of consciousness in nature, it's helpful to look at the way it evolves. Clearly, we're the most intellectually developed species on our planet, with our mind and reasoning ability far outpacing any other creature we share the planet with. But if we look at the development of consciousness from the simplest to the most complex organism, it is not the intellect and reason that come first in terms of evolutionary development, *but sense.*

Even less complex organisms can experience a basic sense of their bodies which allows them to react to their environments. A very simple organism like a jellyfish, for example, can do this to some degree. This appears to be the first thing to manifest itself in the consciousness of the living organism. An ability, however basic at first, to be conscious of the environment through touch, as expressed through its body. *All organisms are possessed of a basic sense of touch, the ability to feel.* This is all that is required to have an experience, even if it is a very rudimentary one compared with ours.

This capacity for feeling begins very simply but can develop into something more complex and nuanced. This occurs even before the intellect or reasoning mind does. Consider dogs, for example. It is obvious to anyone who observes them that dogs have complex emotional lives. They are capable of feeling happy or sad, excited or despondent, hurt or angry. They can be traumatized by experiences that remain stuck in their memories. They can feel nervous and agitated, jealous or envious, and can even feel embarrassed. The emotional life of a dog is complex and varied, all having to do with an extraordinarily complex ability to *feel*.

This doesn't mean that they have an intellect yet, a thinking mind that is developed enough to use reason or logic to solve problems. If such an ability exists in our canine friends, it is relatively limited compared with ours. It is obvious that their senses and ability to feel are much more well-developed. If we look at an organism a little more advanced than the dog however, like a chimpanzee or ape, we see that reasoning and cognitive abilities increase. The capacity to use reason to solve problems in their natural environment is more present. This seems to be the trend. When observing the organisms of nature, feeling comes before thinking in terms of evolutionary development.

There are two more critical aspects of the evolution of consciousness that manifest themselves in more advanced organisms which we can briefly mention. The first is the manifestation of an individual will, an ability to make choices that may go against the collective flow of the community it is part of, and consciousness of the self, which is mostly only visible in humans, but shows itself in more subtle ways in other animals like apes.

When these two elements begin to combine, consciousness of one's own individual identity (however basic this might be in the beginning) coupled with a will that allows it to direct its own actions to some extent, they create the potential choice for the creature to diverge from whatever collective it belongs to. It can "go its own way," so to speak. Within simple organisms, this is not the case. It may be that when an organism is very simple in its development, its consciousness begins more "globally," with the will of the organism being subsumed into the greater will of nature itself. The more it develops individuality, the more conscious it grows of its "self," and the greater capacity it has to diverge from the collective.

What we observe in nature is that as organisms grow in complexity, their consciousness develops the capacity for a more varied spectrum of "feeling," more complex cognitive faculties like intellect or reason, a more clearly defined will that allows them to deviate from whatever collective they are part of, and a greater sense of individual identity and awareness of self. It is clear that if consciousness begins anywhere, it is not in mind or thought but in sense (often the sense of touch as expressed through the awareness of one's own limits or body).

It may be that our bias toward seeing the world as purely physical actually comes from the fact that we use our highest and most advanced faculty to understand it. Because the intellect comes relatively late in terms of evolutionary development, and we see it distinguishes us from other life, we might assume that it is the best tool to use for understanding our position in the world. But to use it at the exclusion of our other faculties is folly. It creates a situation in which the worldview that emerges from it is biased toward its own way of seeing. A perfect order of idealized shapes and forms, whether that is reflected in the nature we sense and experience or not.

We have established that sense comes before mind in terms of evolutionary development, but to properly understand the nature of reality and how our consciousness fits into it, we must make one more important distinction. We must distinguish between "consciousness" and "mind."

CONSCIOUSNESS AND MIND

When we enter into the discussion of a duality between consciousness and the physical world, especially in academic circles, it's usually set up as being between *mind and matter*. Most scientific, philosophical, psychological, and metaphysical literature sets up the duality this way. From this perspective, there's no discernible difference between "consciousness" and "mind." They're basically the same thing. All inner experience and our entire subjective existence are subsumed into the thinking mind. Descartes' popular "I think, therefore I am" is a good example of this, suggesting existence hinges on the ability to think. But if you dig a little, you will find that it is virtually everywhere, and rarely is it questioned. However, if consciousness is *not* equal to mind, if they are different, then this assumption becomes problematic.

In the most general sense, the word mind—as employed by philosophers and physicists—refers to the non-physical space in which the observer resides (or in which conscious experience occurs). What could be termed the "inner" of the subjective observer. While the term mind has generally been used for this space, the word is really a misnomer. It is a broad-sweeping generalization for all aspects of inner experience, whether they are "mental" or not. This is often owed to the assumption that all experience stems from the brain.

We have established that those aspects of reality related to feeling are much closer to experience than thought. Using the word mind obscures this fact and subsumes the whole of living experience into the "thinking mind." We have learned in the evolution of consciousness from simpler to more complex organisms that the ability to sense occurs before thinking.

The senses come first in order of evolutionary development. Therefore if we see "consciousness" as having roots anywhere in nature, it is first in sensory experience.

But there is a level of reality even closer to the root of consciousness than even feeling. An area that comes before all forms, feelings, or thoughts and is foundational to reality itself—and that is *being*. To be means to *exist*. The difference between "consciousness" and "mind" is most easily made visible by considering the difference between *being, the existential level*, and *thinking, the mental level*.

Being is not necessarily mental, though the mind and thought can be part of it. When we consider this, a more apt word than mind to encapsulate the whole of our existence and living experience—thinking, reason, imagination, the body, sense, emotion, etc.—would be *consciousness*. Consciousness is a bigger, more all-encompassing force within which "mind" (and all other experiential qualities) live.

Being is a question of what exists, what has substance in reality and manifests itself in nature. It is neither mental nor material. *Being is first in the causal order. It is primary to thinking.* Descartes' "I think, therefore I am," should really be "I am, therefore I think." You cannot think if you don't exist! Existence, the quality of being, is primary to all other things. To be conscious and to exist are fundamentally connected.

AN EXISTENTIAL CONTINUUM – FROM METAPHYSICAL TO PHYSICAL

We have seen and observed many dualities throughout this book, chief among them the duality between the subject and the object. Together, it is being (consciousness) and materiality (physics) that form reality. One is form and structure, the limits that make experience possible. The other is the subject, the one that is "in the game" so to speak, having an experience and interacting with this matrix. We already know that the material world functions according to laws. We have been studying them for a long time now. What has not always been obvious though, or rather, what we have forgotten, is that the subject is also a part of nature, and therefore should also be governed by laws. The interaction between consciousness and matter is not blind or random. *It is governed by another branch of the natural order.*

It is clear from looking at our world today that an understanding of consciousness is direly needed. All our advances in physical knowledge have not been enough alone to bring harmony to human experience or to human relationships. This is in part due to a hidden bias that a thing is only understandable or obeys structural laws if it is outer and public. Since we all share an inner reality—since it is true of all observers—evaluating those things that are common to experience would reveal much about the mechanisms that govern it, and how they relate to other aspects of reality. It would go a long way to exposing patterns that reveal a greater architecture of reality, the laws governing the cause and effect of subjectivity and its relationship with all other visible entities (both objects and other subjects).

The truth is, consciousness is the "subject" in the subject-object dichotomy. It is the felt and experienced side of duality which gives meaning to the whole display. Consciousness is the self, the experiencer, the observer. I hesitate to say the "I," because as soon as we start talking about "I," we are talking about "ego," which takes us into issues of identity. Consciousness, or the subject, can exist without any concept of itself or sense of identity. This is true for example in the consciousness of animals or young children. Consciousness is a free and independent thing that does not depend on a sense of ego or identity.

Metaphysics and physics are two poles of a spectrum. They belong to the same structure. They both seek the truth, the existent thing, what is real. One side is aimed at explaining reality's physical and non-living elements, while the other is aimed at an explanation of the whole order of reality, including its conscious, living elements. They both aim for objective, universal truths about existence, the laws of nature. But metaphysics also aims to understand subjectivity and the experience of the observer within the same objective lens. Ultimately though—and this is the salient point—they are not separate. This is why neither side could ever achieve a full resolution. Because they aim to understand separately what is not really separate, and hence miss the bridge that connects them.

Physics (the material object) and metaphysics (the immaterial subject) are part of a unified continuum. The truth—the *whole* truth—always encompasses both. Reality is neither a purely material nor a purely spiritual exercise. It is by the sum total of both sides, conscious and physical, that the whole spectacle of created reality comes into being.

CHAPTER 18

BIAS

In order to understand how consciousness fits into the total order of reality, we must first address and eliminate the barriers that stand in the way of introducing it. One such barrier is *bias*. There are two components to this. First there are the biases themselves, which incline us to see reality a certain way (and that often affect us subconsciously). Second, there is the idea that subjectivity is *equal* to bias, that it implies it. The inherent uniqueness to consciousness and subjectivity seems to be at odds with a model of reality that aims for universal truths. That is why we eliminated it from our frameworks in the first place.

The question then becomes, how we can incorporate consciousness without violating the universality of the principles upon which our models are based? To accomplish this, we must understand the nature of bias, how it is different from subjectivity, and how it affects our perception.

BIAS AND PERCEPTION

Perception is a critical and integral part of the nature of reality that is neither well understood nor well defined within the spectrum of the hard sciences. These sciences tend not to deal in perception because it is a subjective quality that is neither material nor physically measurable. Still, perception is an observable phenomenon especially visible in living systems, and is therefore entangled with the nature we study. Because we are conscious beings who experience the world through our individual, subjective lens (even when we do "science"), we must consider perception as a relevant factor in our ability to grasp reality as it is.

Loosely speaking, perception is the way reality is received and interpreted by the subject. It is the way an organism organizes and processes sensory stimuli, influenced by a variety of factors like past experiences, cognitive conditioning, emotions, and environmental context. It is a dynamic interplay between external sensory input and one's internal state, leading to a personalized representation of reality. It is at the very essence of subjectivity and consciousness.

When we enter into the topic of perception, we introduce classes of phenomena that can distort reality for the subjective observer. Distortion is when something is altered or misrepresented from its original form. Consider the optical distortion that occurs when you look at an object through water or when you see the world through glasses that are not your own. Light bends in strange ways that causes the world to appear other than it is. Audio distortion can occur when there is radio interference, causing sound to bend and shift in strange ways. Distortion is a deviation from the medium, from a true state into an altered form. *What is important to realize is that this distortion can also occur at the existential level, with our perception of reality itself.*

One of the most powerful factors that can distort perception for the subjective observer is bias. Bias is a preference for one reality over another. It causes us to "lean" one way or another in our perceptions because we *want* reality to be a certain way. This can be conscious, but more often than not, it is subconscious.

For example, when we have children, we often have an emotional bias that causes us to want to see them in the best light all the time. If they perform poorly in a competition, we may push ourselves to overlook what we see because it goes against the way we *prefer* to see them. When we are children, we may be inclined to block out ways our parents may be deficient, because it is emotionally painful at that point to admit their shortcomings. If we are in the presence of a person we are very attracted to, we may incline ourselves to like the same things they do to generate some commonality between us, or for fear of losing their interest. These biases are often very subtle, working on subconscious levels, causing us to skew reality in one way or another based on emotional factors.

Biases can also occur for psychological reasons. It may be that we have been taught to think in a certain way since we were very young. It may be that we come from a religious family and the ideas they have taught us about the way the world really is can be tied up with our emotional bond with them. Denying the truth of an idea they have taught us may feel like a betrayal of them or our community. Psychological bias may occur because of our schooling. We may have been taught that reality is a certain way by teachers who presented a certain version of things to us. This schooling may be tied up with issues of personal identity where challenging the fundamentals of these ideas means challenging who we are.

Of course here we are speaking only about the experience of one observer, not all observers. In this way we can say that this distortion is a relative phenomenon belonging to the perspective of one subjective entity. But distortion can also occur at the collective level. When most or all members of a group hold the same bias at the same time, distortions can become especially convincing. When reality is distorted at the group level like this, an individual within the group who is able to step out of or clear the distortion will appear to the group to be the one confused! When everyone within a group is under the sway of the same bias at the same time, this distorted perception can appear to be reality as it is, since no one within the group can see outside of it.

We can develop biases for or against individuals, ethnic groups, religions, social classes, political parties, or even sexual or gender identities. But most importantly, we can develop them for or against ideologies, ways of thinking, and theoretical paradigms. These are especially influential, because they affect our perception of reality at the level of truth itself. But regardless of its source, when we are biased, we are not seeing reality as it is. *Bias distorts reality at the existential level, the level of the perception of reality itself.*

A bias is essentially a deviation from "zero," the neutral viewpoint. When you lean one way or the other, when you have allowed an emotional or psychological preference (conscious or subconscious) for one reality over another to take you over, you have become biased. You are no longer at zero. Remember the real number line and how there are positive values on one side and negative values on the other? Bias is a bit like this. You can become positively biased through idolatry and idealization (making something "bigger" or "more" than it actually is), or you can become negatively biased through judgment and condemnation (making something "smaller" or "less"). Either way, as long as you are not looking through the neutral lens, through zero, you are not really seeing reality as it is. Instead, you are seeing a version of reality that is distorted, in equal proportion to the strength of your bias in either direction.

BIAS IN SCIENCE

You may be thinking, okay this whole bias thing is well and good. I can see how we can develop these various biases throughout our lives, and how that can distort our perception of reality based on the ways we prefer to see things. But surely this can't happen in science. After all, science makes a point not to do this. The rule of maintaining total objectivity in all things is meant to prevent this very thing. Scientists don't make opinions, they deal in facts based on empirically verifiable evidence. That's what makes science foolproof.

Yet even a scientist can develop biases. First and foremost, *because a purely objective viewpoint is impossible*. The scientist is a subject, even if they train themselves toward objectivity. The observer and their perception are unavoidably involved in every observation. We can work to reduce biases. We can make an effort and intend to see things neutrally, but the idea that we can achieve "total objectivity" is just another illusion.

The idea this is possible is a reflection of its bias. The material realist bias, the idea that reality is fundamentally physical and that an understanding of physical law will one day be able to explain everything, is the most widespread and common psychological bias within the scientific community. This was originally created because subjectivity, by its very nature implies an individual perspective, and science is looking for the universal. Science aims for an understanding of reality as it is, irrespective of personal opinions or perspectives.

Yet the issue was never really with subjectivity. *It was with the biases that we assume come along with it.* We didn't want personal opinions to get mixed up with our objective analysis. We wanted to see things in a way that was irrespective of the personal or subjective. This is very understandable. The striving for objectivity, an impersonal and neutral perspective, ensures that we are seeing things as they are. But the idea that this same neutral view cannot be turned *on the subject itself*, was overlooked.

Subjectivity and bias are not the same thing. It is true that the word subjective implies a relative perspective, the viewpoint of one entity and not a universal one. But it is arriving at the universal from the personal that is the basis of all science—even the way we do it now. Science always begins and ends in the personal. This is unavoidable. Our true goal is not really the elimination of the subject, but the reduction or elimination of bias.

Science, if we want to think of it in its broadest sense, is the unbiased observation of nature to discern patterns and laws. That this should be impossible for subjectivity is not a given. The key word is "unbiased." If we want our study of nature to extend to the subject, how can we do so without our biases getting involved?

NEUTRALITY AND TRUTH

In order to incorporate consciousness into our ontological models, we must first aim for objectivity. This means our ability to get away from the special viewpoint of the individual observer and see things in a way that is more neutral and general. Not from our individual lens but from a universal one. If moving from a subjective to an objective viewpoint was not possible, science as we know it would also not exist. So our first step is to strive for objectivity.

Secondly, we must disentangle the idea of subjectivity from bias. They are different things. We must realize that subjectivity is as much a part of nature as objectivity, and that including it in our models is necessary for a whole accounting of things. We must strive to observe subjectivity from an objective lens. We can use the same methodology of moving into a neutral and objective viewpoint for the observation of subjective phenomena as we do with physical ones. Ideally we examine them together in the same bucket, to identify patterns and discern laws.

Finally, we must identify and eliminate our biases. These subtle inclinations that distort our perception of reality and cause us to want to see things in one way over another. These biases can come from many places, including our pre-established modes of thought and methods of analysis. We must be willing to step back from the way we are doing things, to be *neutral* enough to see patterns that exist at higher ontological levels. The dissolution of bias ensures the elimination of distortions that cause us to see reality other than it is and to emphasize pre-established ways of doing things that may come from subjective preferences.

Whether social and cultural, spiritual and religious, or academic and philosophical, our established frameworks of thought and ways of doing things persist and cycle through generations. The dominant ideological and theoretical paradigms of our time get taught to our children who absorb them uncritically. Therefore it is often the case that we see the world the way we do simply because the ones who came before us did. It is precisely those paradigms that are most foundational to our psychology, often rooted in our original thoughts and teachers at early points in our history, that are the most important to identify and evaluate. *Our root beliefs about the nature of reality are the hardest ones to see outside of and to change.*

Still, it is possible. Science is the process of observing nature through an unbiased lens, identifying patterns, and drawing conclusions. The road to truth is this simple. Reality is observable. This is achievable for the entirety of our reality, subjective and objective, inner and outer. The goal is to look from a viewpoint that is neither positively nor negatively charged, but *neutral*. This is what it means to be *"unbiased."* It is the road to truth, whether within our personal lives, or within our science. By letting go of our biases and assumptions with regard to the nature of reality, we can open a road to understanding the laws of nature as they pertain to the whole, the reality of consciousness and the reality of physics.

CHAPTER 19

BELIEF

We have established that there are many factors involved with consciousness that are absent on the side of the material object. Chief among these is perception, the individual way in which reality is received and interpreted by each observer. Given that perception is different in every case and our science is generally looking for the universal, it difficult to quantify or incorporate into existing frameworks. To understand this, we must return to one of the most central and important concepts of the twentieth century—*relativity*.

THE RELATIVITY OF CONSCIOUSNESS

Relativity is generally understood within the context of physics. While we previously believed that space and time were absolute constructs—the same for all observers—we learned with Einstein's revelations that they were in fact variable. Both space and time were dependent on the frame of reference in which we took our measurements, different for every observer.

There are some ideas implicit in relativity that are central to an understanding of consciousness too, though Einstein and others within the physics community who were more focused on the physical implications of these discoveries did not fully realize or comment on them. First is the inherent nature of a "frame of reference." A frame of reference is another way of saying a "perspective," "vantage point," or "way of looking."

What is inherent to a frame of reference that is not said explicitly in the theory of relativity, is that it must be *chosen*. It is not an objective feature of physical systems that exists somewhere out there on its own. In fact, it is not physically measurable at all. What we can witness and verify empirically is that it is always dependent on an observer. What is implied by a frame of reference that we don't say out loud, is *consciousness*.

All consciousness—even in a basic lifeform like an amoeba—is a unique position from which to view the whole. Relative means particular to one way of looking, not universal. *This relativity is an essential quality of consciousness.* We established that one of the principal differences between organic and inorganic systems is that organisms have an element of in-dividuality that is absent from inorganic substances. Living systems seem to exhibit a quality of inward integrity, wholeness, and individuality that is absent from non-living ones. These elements are difficult to fit into our models given that they seem to suggest a sort of infinite variability, and our models deal in universals.

Here we begin to get to the heart of the matter. Relativity is an essen-tial attribute of consciousness. It leads to individuality, which is the nat-ural result of each entity's unique frame of reference with which to view the whole. This relativity and individuality, embodied by the conscious observer, serves as the subjective half of the subject-object duality. *The subject ensures there is always a frame of reference.* A way of looking that is not universal, but relative. This dynamic is essential to reality itself. It is why even physics cannot come to a complete description of a physical system without accounting, in some way or another, for the individual point of view of the one doing the measurement.

This relative perspective stands in opposition to the generalized view-point of the intellect, which is neutral, uninvolved, and objective. These "subjective" and "objective" viewpoints form a polarity. They are two frames of reference from which to view the whole. One looking from the inside out, and the other from the outside in. Only through their sum total can the full spectrum of reality be known and understood.

RELATIVE REALITY

When we consider the relativity of consciousness, that each observer ex-periences the world from their own unique frame of reference, and we add to this the immense variety of experiences that accumulate in memory and later affect behavior, the individuality we observe in nature makes more sense. Every life, no matter how simple, is a unique expression. It is this individuality that serves as the opposite pole to the universal and objective elements we tend to seek in the sciences.

In considering that every consciousness experiences the world from its own relative perspective, it is worth considering what mechanisms contribute to creating this inner experience for the observer. We have already identified a few. We know that perception is paramount and that it can be affected by a variety of factors, including the state of the body, cognitive biases (whether psychological or emotional), the environment, and others. Perception ensures that the relative reality of the observer is always unique and variable.

We also know that many of the living organisms on our planet do not possess cognitive faculties as advanced as humans, preventing them from interpreting the world through complex logical frameworks. While this gives us a leg up in our ability to analyze the world in certain ways (especially objectively), it is not foolproof. We know that errors in logic can occur which can distort our perception. We can become convinced of the truth of something simply because those before us saw things a certain way. Or our past experiences can convince us of the truth of something, even if current evidence in our environment does not support it.

All of this is to suggest that reality itself is relative. Of course, this is not *all* reality is. There is also a universal, objective element to it too. Our inner experiences occur within a physical matrix that behaves according to predictable laws. But what we can observe is that these two elements coexist and interact with each other. It is only through their combination, the relative and the universal, that the created reality we experience through our senses comes to be.

We are familiar with the objective side for the most part. This is mostly what we look at in the sciences—especially physics. The relative reality we experience through our senses however, is not as well understood in the context of nature's laws. We have already identified some of its defining features. We understand that consciousness always has an individual component to it that is essential to its nature. We understand that this consciousness is dependent on perception which can be altered by a variety of factors. We understand also that this consciousness is at the heart of the nature of relativity. But there is one more factor involved that ties together many of these threads. Consciousness, relative reality, and the nature of truth are all bound up with each other by one more factor that potentially bears the greatest influence of all over our experience of the world—and that is *belief*.

BELIEF

Belief is the conviction or acceptance that something exists. Believing something means that, from our perspective, it is real. Just as we have established that there is a relative reality and a universal reality, a level that pertains to the observer and another that is general, truth is also possessed of this same duality. *There is an objective truth, that which is true irrespective of subjectivity or any individual perspective, and there is relative truth, that which is true for one observer and no one else.* We must remember that these are not *separate* realities, but one reality expressed in two ways.

To believe something means *to declare it is true*. A declaration of truth is an existential act. You are making a statement about reality itself—that something exists. The more conviction with which you declare this, the more full and definitive becomes its existential quality within your relative space. In other words, by declaring something to be real, *you make it real for you. That which we invest with belief becomes our relative truth, the reality of our experience.*

To be clear, this doesn't affect objective reality. Our influence over the general, shared space is relative. But influence over our own is not. The depth of our conviction about something, how convinced we are of the truth of it, determines the degree to which it manifests in our own relative experience. In this way, *our relative experience is literally created by our truth about it.* Belief is one of the greatest creative powers in the universe. It just happens to live on the side of the observer. It is also why an examination of our beliefs, of our *truths*, is so important.

CONFUSION

Of course, our relative truth and the objective truth do not always line up with each other. It could be that our interpretation of reality, being skewed by a variety of things, including past experiences or biases, could create the impression that reality is other than it is. When our relative reality and objective reality are out of step with each other, we experience *confusion*.

Confusion is a distortion of reality at the existential level rooted in a misinterpretation or misunderstanding of objective truth (or "natural laws"). This can have a significant effect on our experience. If we believe, for example, with a great degree of conviction, that there is a ghost in the room, we will experience its reality. If we believe deeply that there is a devil down below that is hunting us, we will experience the reality of it. A lack of belief in our own value, the belief that we are worthless, can cause us to accept circumstances that are undignified or harmful. Belief that we deserve more than others can cause us to cheat or steal. Belief that we are victims can cause us to justify great harm to others. Belief that our existence will terminate at the end of our lives has the potential to create great existential dread. When we believe something, we create the experience for ourselves. In this way, *any illusion can be made real, and any truth can be made into an illusion.* The creative power of belief ensures that we can experience, within our own relativity, literally anything. This is what makes an evaluation of our beliefs—of our *truth*—so important.

Of course, this is as true at the communal level as it is at the individual one. When many people within the same community hold the same belief about something at the same time, especially if it is held with deep conviction, the perception of reality for that group can become greatly affected. By putting things this way, it may sound more innocuous than it actually is. This distorted perception can lead to severe and sometimes devastating consequences.

In the fifteenth to eighteenth centuries, it was so commonly believed that there were witches among us who communed with the devil, and so deeply we were convinced of the truth of this, that we burned people alive and cheered as they screamed. So many groups throughout history have been convinced of their ethnic or genetic superiority over others that millions upon millions have been murdered with incredible brutality. Beliefs shape our relative reality at both individual and collective levels. *The importance of this cannot be overstated.* Therefore, an honest examination of our beliefs is paramount if we hope to bring resolution to some of the most dire problems of our existence.

Individually, these beliefs can have enormous power. Together, even more so. Just like logic, beliefs can exist on a tree, where one "truth" can lead to belief in another, and so on. That they can work together and reinforce each other means that it is crucial to evaluate them not only individually, but also as a system.

BELIEF SYSTEMS

Our metaphysics is our assessment of truth. It is the network of existential statements we have made internally about what is real. We may not know it by this word or by this name. We may simply call it our belief system, spiritual practice, or religion. But what is important to realize is that whether you subscribe to spiritual concepts or not, whether you have ever set foot in a temple, synagogue, church, or mosque, or whether you have spent two seconds thinking about your core beliefs about reality, you will still have come to some conclusions about it. You cannot help but have a metaphysics because you are conscious. Because you are having an experience and receiving information about the world at all times, you must have some way of determining what is true and real about it and what is not. The system of beliefs which formulates these interpretations could be called your "belief system." Your belief system is the single most influential factor in creating your relative reality.

A belief system is a *set* of beliefs—with connected logical or existential statements—that work together to paint a more holistic picture of reality. These can come from culture (the relative truths of a group of people from a particular place, like a city, country, continent, or tribe). They can come from family (including those of our parents, extended family, or their ancestors). They can come especially from spiritual or religious ideologies, which often make direct statements about what reality is or isn't. But regardless of the source, it is our belief systems that most affect our perception, filtering reality through our ideas about what it should be and presenting a version of things that reflects our relative truth. Potential confusion within these belief systems necessitates a set of tools with which to interpret them, to determine the nature of the *objective* truth and whether our relative truths are in line with them.

THE PRACTICAL USE OF PHILOSOPHY

Historically, the field that investigated the truth was called philosophy. All other branches of knowledge dealing with truth—including metaphysics and science—were considered part of it. At its core, philosophy is an assessment of truth, what is *real*. Definitions that go beyond this run the risk of complicating what is not meant to be complicated. The goal is to reveal the true nature of things by whatever means we have available to do so. When viewed in this light, philosophy is not really an academic "field" perse, but a ubiquitous human endeavor, for we cannot live without interpreting the world through some set of beliefs, and we cannot act without this system informing in some way the way in which we do so.

There are some things that, by one avenue or another, we believe are real. We cannot help but have made these assertions, because by virtue of our being in the world, we must have a way of interpreting all the sensory and intellectual data we receive about it.

Philosophy, in its purest form, is simply the act of developing an intentional system for the interpretation of this data. It is a way of assessing our relative truth and ensuring it lines up with the objective one. Since our beliefs are so consequential in the reality we experience, the true benefit of philosophy is to transform our experience through an honest examination of our beliefs, and through reason, intuition, or whatever other tools we have at our disposal, to discern truth (reality) from illusion (misunderstanding).

TRUTH AND REALITY

The discernment of truth from illusion is the primary work of philosophy. By assessing our interpretation of reality on both subjective and objective levels, when we can look at the whole of our ontological reality within the same philosophical lens, we open a road to resolution. Reality is an observable entity with discernible properties. When we understand that this is just as true in the realm of subjective experience as it is within the "objective" space we share with each other, we will make progress. This is as true for our individual lives as it is for our science.

The objective side of the subject-object duality is the side that deals in *laws*. It is the structure within which all relative reality exists and expresses itself. While we have explored this realm to a great degree with respect to its physical structures, we are completely clueless when it comes to the architecture of experience. We have not sufficiently explored the patterns and mechanisms visible in our inner space. An understanding of natural law as it pertains to this space will give us the tools we need to finally accord relative truth with objective laws, bringing order to the side of reality where we have so consistently manifested chaos.

Philosophy—or rather, metaphysics—is not an empty academic exercise. It is the most profound method we have to change our reality by reframing the truth and therefore transforming our experience. Our same ordinary reality can present itself in a new light, not because it has changed, but because we have. The truth has the power to liberate us from false beliefs and alter our perception. By clearing misunderstandings from our consciousness, our experience of it can transform. Truth is medicine for the mind, a balm for the soul. Our intentional and conscious liberation from false beliefs, from a misunderstanding of natural law, provides a road to clarity. This clarity is peace.

CHAPTER 20

TRUTH

The word "truth" is often avoided in professional circles. We may say things like the study of knowledge, epistemology, philosophy, or science, but rarely do we say—in an official or professional capacity—that we are studying *truth*. Moreover, it is often the case in science (especially in fields like physics) that we say our inquiry is into "the physical world," "physical reality," or "the external world," but rarely do we say that we're studying the nature of *reality*. Yet in the end, it is both the nature of *truth* and the nature of *reality* that are at the heart of our most important questions.

You often see people bend over backwards in academic and scientific circles not to use this word. You may hear "reality as it is," "reality as it really is," "reality per-se," "the real," "what is," "all that is," "things as they really are," etc.—all different ways of saying the same thing. That there is a truth, a *reality*. It is the discernment of the nature of this reality that has inspired millions throughout history to explore work in the fields we now call philosophy and science, and to a degree, theology, spirituality, and religion.

TRUTH AND FALSEHOOD

Part of the reason professionals within the former fields tend to avoid using this word is for fear it will be associated with religion. The truth, it seems, is a phenomenon that has been owned wholesale by religion for the better part of our history. In reality though, truth is not something that inherently belongs to religion or outside of the realm of everyday experience. *Truth is the nature of reality itself.* It is the existent thing, that which *is*. Difficulty in disentangling truth from "God" or religion keeps many from seeing its fundamental entanglement with science, the study of nature, and its relevance to our everyday lives.

There is something important to realize about truth. If there is a reality, a real thing, then there is also an *unreality*, an unreal thing. Understanding the nature of this "unreality" is an important part of metaphysics. It has gone by many names in many places. The ancient Persians used to call it *druj*, often translated as "lie" or "deception." It represented the diametric opposite to truth. These two formed a duality—that which is true (*asha*) and that which is untrue (*druj*). Yet regardless of the terminology, the implication is clear. That there is an opposite to truth. An antithetical. In purest terms, we can simply call it *falsehood*.

We established that truth has two forms, one relative and the other universal. Relative truth is what is true for one observer and no one else. Universal truth is what is true in general, for all observers. Falsehood is similarly dual. There are things that are false for one observer, with regard to their relative reality, and things that are false in a more universal sense, for all observers.

Falsehood is key to understanding truth because it is its ontological opposite, and generally not as well understood. So, let's explore it. First and foremost, what is false is *not real*. We can call it an "illusion." Yet we have to make one more important distinction here to properly understand this. In saying that falsehood is not real, it may be tempting to consider that this means it doesn't exist. That is not true. *Illusions do exist.* They simply exist *as* illusions.

This is the crux of understanding them. Unreality exists (it has ontological status) and does not exist (it is unreal) at the same time. *Just because something is unreal, that doesn't mean we can't experience it.* We can experience an illusion and be conscious of the fact that it is unreal (like watching a movie), or we can be so deeply in it that we confuse it with reality. Remember that we can also bring to life any illusion by simply believing in it with deep conviction. *Truth and falsehood together form a basic duality through which our relative reality is defined.*

To be clear, we are not assigning any moral value or judgments to this. *We are not talking about good and evil or right and wrong.* We are pointing to an ontological duality that is part of the construct of reality itself, and through which our relative reality is defined. There are real things and unreal things that together give definition to our experiences. Assignments of moral value above and beyond this are extraneous.

Therefore when we say "false," we do not mean "wrong" (especially in a moral or logical sense). Instead we mean "illusory" or "artificial." Not real. It is like a false wall. A false wall gives the *impression* of a wall. It looks and behaves like a wall. But in truth it "*is not.*" This is the meaning of this side of the duality. The false wall exists. We can see it, touch it, and interact with it. But it is an illusion. A trick. A "lie" if you want (if you can disconnect this from any moral or ethical undertones). That is the nature of the false, ontological entities that exist as the antithetical to true or real things.

Reality is a spectrum of interrelationships between true and false, *that which is ("reality")*, and *that which is not ("unreality")*. A lot of suffering in the world comes from the misinterpretation or misunderstanding of truth—the belief that something false is true (*that an illusion is real*) or the belief that something true is false (*that some reality is an illusion*). This is the essence of confusion. It is the way illusions are made real within our relative experience.

Resolving this confusion in our relative reality is the road to clarity. This process begins with a discernment of natural laws (objective truths), the way reality really works, and a subsequent alignment of our relative truth (our beliefs) with those laws. This alignment clears conflict between inner (subjective) and outer (objective) aspects of reality. The relative and universal begin to dance together and cooperate with each other. Clarity (clear-seeing) is the ultimate goal of metaphysics. It has at its heart *what is*, the nature of being.

At the highest level, it is the concept of being that is the common denominator between all fields dealing with truth: philosophy, science, religion, spirituality, theology, etc. Things either are or are not. When we use this word in language, we are referring to an existential quality, that something exists. It is an easy way to see how consciousness (a being) and truth (to be) are connected.

Being speaks to the fundamental question at the heart of metaphysics. *What is real?* This is the meeting point of all the previously mentioned fields dealing with truth. It is where philosophy and physics meet. Where the measurable and the immeasurable meet. Where the subject and the object meet. In the nature of being, we find the aim of all philosophical and scientific pursuits, the search for truth, what is *real*.

CONCEPT AND EXPERIENCE

To truly grasp what is real, we must make one more distinction. We must differentiate between *concepts*—ideas or theories about reality—and *experience*—our firsthand contact with it. These are two streams of information coming from different sources, one from the mind and the other from the senses. A concept is a mental entity. It takes shape as a thought-form. An experience is personal and direct. It comes from contact between the subject and the world around them. In other words, concepts are mentalizations of reality while experiences come from direct contact with it.

Concepts are impersonal. They are detached to some degree from the person that we are. While concepts can be about us (related to our own experience) or more universal (related to the nature of reality as a whole) the concept itself always remains in an objective space, removed from our direct experience. This is why conceptual knowledge is not the same as experiential knowledge. If we want to enact change in the real world, concepts have to be applied at the personal, subjective level. Otherwise, they simply remain concepts.

We make an effort in science to remain as objective as possible. This has served us well in many ways, but there are consequences to looking at things this way that we don't always consider. When we remain objective *too* often, ironically, *we can lose touch with reality*. The real world—not the one known through concepts—comes to us through the senses. This is the essence of empiricism which forms part of the scientific method. When everything remains at the impersonal and general level, we lose touch with the actual world around us. Given that the being we are is known through experiences delivered to us by the senses, this essentially means that *we lose touch with ourselves*. This is why subjectivity matters, because regardless of what the objective truth might be, it doesn't mean much if we are out of touch with what is actually happening within our own experience.

A disconnect between concept and experience can occur for many reasons. For example it can occur in fields that require a great deal of conceptualizing subjective data, like with therapists or scientists. Our perception of things ends up becoming very conceptually driven, which can put us out of touch with the world as it presents itself to us empirically. We can become entranced with general trends and patterns and lose touch with the nuance that is always present in the real world. For example, a therapist overly reliant on conceptual knowledge can potentially misunderstand or overlook the particularity of their patient's personal situation. They may try to fit them into the mental schemas they have been taught, rather than to take their situation uniquely. A scientist may get so engrossed in a theory that they miss empirical evidence in front of them, the individual nuance of a situation. This can occur in virtually any field, where conceptual schemas of the mind get the upper hand, and imperfection and nuance that are inherent to the real world get overlooked in favor of mental conceptualizations and frameworks.

This disconnect often even occurs on a personal level. The conceptual space can serve as a kind of refuge from having to be honest about our own subjective situation. We can use it as a safe haven—sometimes reflexively when we are confronted with something that frightens us or makes us uncomfortable—to retreat into concepts and rationalize everything. This creates distance between us and the thing that scares us, giving us an artificial feeling of safety. But of course when we do this, we haven't truly solved the problem, since our underlying feelings are still there and we haven't dealt with or confronted their reality (which exists at the level of our subjectivity).

Furthermore, there are moments when our mind may override what we are feeling with comments or suggestions like: "This is the incorrect emotion. I should really be feeling {blank}" or "I'm merely experiencing {such and such} condition." All of which puts us out of touch with what we're actually feeling and puts us back into an objective space of ideas and concepts that are general. Being objective can mean not having to be "you." It can be a tempting refuge when our experiences are bad or overwhelming.

In the duality between concept and experience, it is the intellect that deals primarily with concepts and the senses that deal primarily with experience. This is why concepts are more associated with thought and experiences with feelings. This relationship hints at another dual aspect of truth. It has an impersonal, conceptual aspect to it, and a personal, experiential one. These are mirrors for objectivity and subjectivity. By their sum total our reality takes shape.

There is often the assumption that philosophy and metaphysics are based purely in concepts and can't be applied to experience. But this is not true. In fact, it is the reversible interaction between concepts and experiences, that concepts can affect experiences and vice versa, that creates the type of dynamism that makes philosophy useful. When our perspective changes, our relative truth changes too. When our truth changes, so does our experience.

In the final analysis, all assessments of reality begin and end in experience. Concepts give us an opportunity to interact with things in a more freeform space liberated from direct experience. They give us an opportunity for objectivity. It is an incredibly useful tool, but to consider it "reality" would be a mistake. It is grounding in the perspective of the observer, a center from which to view the whole, which gives reality its meaning.

THE CENTER OF REALITY

Relativity has a central and important meaning in metaphysics. It implies a *point of origin*—a relative center within which all reality occurs. From this point, from our own frame of reference, *all things are relative.* This frame means being in a specific point in space and time with which to look back on the whole. It is required to make sense of any picture of reality—even in physics. Aside from Einstein's introduction of the necessity of this frame for the accuracy of physical measurements, there is also the fact that all scientific analyses begin and end in the empirical (personal, sensory). Since it is present at the beginning and end of every process of our science, we must consider what it means in and of itself in our ability to grasp reality as it is.

One thing to consider is that if our only means to understand empiricism is through empirical methods, we run into a problem of circularity. What other method is there to understand experience except through experience itself? This seems to suggest that *experience is fundamental to reality.* We can conceptualize about experience from within the freeform space of the mind, but even this occurs in experience. If reality begins anywhere that is discernible to us, it is at its relative center, *in the frame of reference of the observer.*

This hints at a central and important truth—*reality is defined relative to the observer.* This is at the heart of metaphysics. The hard sciences approach this problem from the opposite direction, defining the observer in relation to the material environment. This is an expected outcome given the emphasis on objectivity, quantification, and intellectually-driven conceptual frameworks. Yet when we enter the domain of living things with all their inherent qualities—memory, learning, adaptive behavior, evolution, qualitative data, inner experience, individuality, etc.—this framework falls apart. It is inadequate to handle all of the data.

This is the primary problem. If we hope to understand reality as a whole, we must have a system that can accommodate all of the data we receive about it. This includes information from all streams of knowledge: conceptual and empirical, quantitative and qualitative, relative and universal. We require a system possessed of a singular tree of connected logic that can accommodate it all, producing a vision of reality that is whole.

THE WHOLE ORDER OF REALITY

To build a system that can accommodate all of the data, we must first create a way of accounting for everything. One way is to consider reality from the perspective of six basic questions or angles (used in certain fields as a way of understanding a situation comprehensively): the who, what, where, when, why, and how. The ontology that emerges from the hard sciences answers four out of six, but two are left out entirely. We can consider "what" to be the object, "where" to be in space, "when" to be in time, and "how" to be by way of the laws of physics (as described through mathematical models). Essentially we have material objects (what) in space (where) and time (when) whose motions are described by the laws of physics (how).

What have we left out? First there is "who" (which can be considered the subject), and finally "why," which we can call reason (or philosophy). Of course when we reintroduce "who" into the picture, "how" must then also include the laws explaining the causal interactions between the subject and all other entities (whether objects or other subjects). "What" (the object) would then also need to expand to include entities that are not strictly physical in nature, such as mental entities like "ideas."

In our current assessment of reality and the model that emerges from it we focus primarily on the what, where, when, and how, but leave out the who and why entirely. The latter two are not considered to be fundamental to understanding nature. This results in a system that can describe objects and their behaviors very well, but is silent when it comes to their subjective relevance. It does not sufficiently consider that the experience of reality is something that occurs purely in subjectivity, and that this must be entangled somehow with the rest of our observed reality. It stopped looking at the "who." Similarly, it eventually stopped caring about why.

Why should there be such a thing as a relative reality? Why is it different for every observer? What is the purpose of this experience? What is the end goal? These questions matter. Not just for ourselves personally, but also for our science. Especially when we consider that our experiences and relativity are so bound up with the nature we study, does an answer to the question why not seem key to understanding the whole?

WHY?—A REASON FOR BEING

Ultimately, reality has a purpose. There is a *why*. And it has everything to do with us—the observer. Life is about more than just objects in motion. It is about *experience*. This experience has more significance within the cosmic order than we imagine. We are used to thinking of ourselves as so small. Just cogs in a giant machine that chugs along whether we're here or not. But what if I told you that wasn't true? What if I told you that *you*, the specific individual that you are, were *essential* to reality? That nothing could exist without you?

Shifting our ontological outlook from a worldview based in physics to one based in metaphysics has two primary implications. First, it implies a shift from a parts-first analysis of reality (physics) to a holistic treatment of reality (metaphysics). Second, it is indicative of a shift from a focus on *experimental* outcomes, to a focus on *experiential* outcomes. This is the main point. When the observer reenters the conversation and is understood to be essential to the cosmic order, the meaning and purpose of reality comes more into focus.

At the end of the day, it is not really a *thing* we are after. It is an *experience*. There is no physical object (or understanding of them) that can or will deliver us from the chaos and suffering in the world, whether inside of ourselves or in the world at large. What we are really looking for is a *feeling*. A *goodness* within (in our feeling and experience, about ourselves, about others, and for the world at large). We want to feel this way consistently, in a way that is maintainable in the face of any change.

The Greeks had a word for this. They called it *eudaimonia* (εὐδαιμονία), meaning happiness, welfare, or human flourishing. *This* is what we're really after, by whatever means we have available to achieve it. We typically think of this eudaimonia in physical, three-dimensional terms. In a surface-level way. We imagine having all the *things* we think we need to experience it. An abundance of food or resources. Clean streets. Organized societies. That people *look* healthy. *But this is not it.* What we cannot see with our physical eyes that is much more relevant to the question is how people *feel*. It is not a physical thing, this happiness and welfare. It is an *experience*.

So long as we remain fixated on physical appearances and outward forms, we will miss the true reality that lives underneath the surface, where the point and purpose of this whole creation finds its mark. That we might experience *paradise*. A deep, abiding, and sustainable feeling of goodness, fullness, and plenitude within. Harmony between all our inward parts (mind, heart, and body). Peace. Wholeness. For ourselves of course, but also to emanate such a thing into the world and share it with everything and everyone that surrounds us. *This* is the ultimate point of life. It is the true goal of all this metaphysical work. Joy—a living paradise—*that* is our reason for being.

The basic dualities we have identified in this book are a good start. They have begun to open the door to a more comprehensive science of energy. One in which we, as subjective observers, are included and accounted for in the interplay of the cosmos. This framework creates a bridge that can connect the conscious realm of experience with the physical realm of objects and motion in a profound way. Most importantly, it opens the door to a deeper and more comprehensive understanding of natural law. These laws are our salvation.

By understanding the nature of reality, the laws that govern our experience, we pave the way to a resolution to our inner chaos and suffering. By according ourselves with these objective truths as they relate to the functioning of the universe from within the perspective of the one who looks, the subject of cognizance, the one who is at the relative center of reality, we can pave the way to a more peaceful and harmonious experience, and a flourishing of human society.

AFTERWORD

This is the first book in a series. This series lays out a holistic metaphysical ontology that accounts for all major phenomena, including our own consciousness, physics, notions of virtue, and a path to enlightenment. It took ten years to create that system, and another two years to write this book.

It has been my intention for a long time to discover a true explanation for the role of consciousness in reality. Something that can explain, in clear, technical terms, how it interacts with the matrix of the material world. When I was a child, my mother used to drop me off at the local bookstore or library and I would spend hours in the Metaphysics section looking for this type of explanation. As it turns out, I ended up writing the book child-me was always looking for.

With that said, our journey is not over. Since the ideas presented in this book series challenge convention, I needed to take the time to lay out some groundwork. This book serves as something of a preparation for what is to come. I consider us to be somewhat lost when it comes to the feeling and experiential aspects of life. The social, emotional, and mental chaos in the world is bordering on catastrophic. Book 1 comes and finds us wherever we are lost in that jungle. It takes your hand and brings you to the foot of a bridge.

This bridge is the groundwork that can transition us to a new way of looking, an ontology that bridges the material and immaterial. If Book 1 finds you where you are and brings you to the bridge, Book 2 builds it. It lays out the metaphysical framework within which the whole can be understood. It is a necessary prerequisite for Book 3, which is the book I was always trying to write. In that book, we walk across.

The nature of my inquiry has always been geared toward understanding and explaining phenomena for which we do not have clear explanations, like the nature of emotion, consciousness, the ego or inner child, and virtues held in such high esteem like compassion, patience, or trust. I have a technical mind, so rather than understand these things esoterically, I was more interested in understanding them *technically*, as a part of the fabric of reality.

In other words, I was interested in learning *natural laws*. Laws of nature that one can rely on as a solid foundation in understanding one's own relationship with the world around them. Right now, we don't really have laws. We have a lot of opinions, and that contributes greatly to the fragmentation we have between us, as well as a lack of security we often have about our own beliefs.

I was looking for a clear understanding of inner aspects of reality based on a reliable science I could lay out to a layperson. An explanation anybody could understand. That is what motivated this book and the others that are coming. It was out of a sincere desire to be helpful. The importance of areas of life dealing with my own experience and the parts of reality I could not see with my eyes stood out to me as important or even essential. At the same time, I was always fascinated by science and enjoyed the sure, discrete answers it would provide.

When I began this process, a thought occurred to me. If I am going to be able to divine these answers, they will be found straddling the line between reason and intuition, between thought and sense. It just seemed intuitive that it should be that way. That if there is going to be an answer to the question of "reality," what it really is and how it really works, it is going to have to use both sides of the brain in equal proportion.

Ten years (and seven notebooks) later, countless individual realizations eventually built into a network of interrelated ideas revealing a grand architecture. A science of energy that transcends mere physicality and is able to speak to the whole creation. I realize how big that sounds—but I don't think it is. I think it is only because we're accustomed to thinking of reality in a certain way that it sounds so grandiose. The reality is, the truth is always there to find, there to observe and to discover. If we are brave, we can find it. We need only look.

NOTES

CHAPTER 6: THE SUBJECT AND THE OBJECT

1. Erwin Schrödinger, Nature and the Greeks, Cambridge University Press, Cambridge, England, 1954, pp. 53-54.
2. Albert Einstein, "Ernst Mach," Physikalische Zeitschrift 17 (1916): 101-102.
3. Albert Einstein, "Physics and Reality," Journal of the Franklin Institute, Vol. 221, No. 3, March 1936, p. 351.

CHAPTER 7: THE OBSERVER

1. James Clerk Maxwell, "A Dynamical Theory of the Electromagnetic Field," Philosophical Transactions of the Royal Society of London, Vol. 155 (1865), p. 466.
2. Walter Heitler, "The Departure From Classical Thought in Modern Physics," in Albert Einstein: Philosopher-Scientist, vol. 1, Paul Schilpp (ed.), Harper Torchbooks, New York, Harper & Row, 1949, pp. 181-182.
3. Werner Heisenberg, Physics and Philosophy, HarperCollins Publishers, New York, 1958, p. VIII.
4. Werner Heisenberg, Physics and Philosophy, HarperCollins Publishers, New York, 1958, p. XI.
5. Erwin Schrödinger, Science and Humanism, Cambridge University Press, Cambridge, England, 1961, pp. 154-155.

6. Albert Einstein, Boris Podolsky, Nathan Rosen, "Can Quantum-Mechanical Description of Physical Reality Be Considered Complete?" Physical Review, 47, 1935, p. 777.
7. John Bell, "On the Einstein Podolsky Rosen Paradox," Physics Physique Физика, 1, 1964, pp. 195–200.
8. Fritjof Capra, The Tao of Physics, Shambhala Publications, Inc., Colorado, 1975, p. 68.
9. Werner Heisenberg, Physics and Philosophy, HarperCollins Publishers, New York, 1958, p. 103.
10. P.W. Bridgman, "Einstein's Theories and the Operational Point of View," in Albert Einstein: Philosopher-Scientist, vol. 1, Paul Schilpp (ed.), Harper Torchbooks, New York, Harper & Row, 1949, p. 349.
11. Albert Einstein, "Autobiographical Notes," in Albert Einstein: Philosopher-Scientist, vol. 1, Paul Schilpp (ed.), Harper Torchbooks, New York, Harper & Row, 1949, pp. 11-13.

CHAPTER 8: MIND AND MATTER

1. Max Planck, "Das Wesen der Materie" (The Nature of Matter), speech at Florence, Italy, 1944.
2. Erwin Schrödinger, What Is Life? With Mind and Matter, Cambridge University Press, Cambridge, England, 1944, p. 119.
3. Sir Charles Sherrington, Man On His Nature, Cambridge University Press, Cambridge, England, 1949, p. 200.
4. Sir Charles Sherrington, Man On His Nature, Cambridge University Press, Cambridge, England, 1949, p. 247.
5. Bernard d'Espagnat, On Physics and Philosophy, Princeton University Press, New Jersey, 2006, p. 412.
6. Sir Charles Sherrington, Man On His Nature, Cambridge University Press, Cambridge, England, 1949, p. 250.

CHAPTER 9: FEELING AND THE SENSE OF TOUCH

1. Albert Einstein, "On the Method of Theoretical Physics," The Herbert Spencer Lecture, delivered at Oxford, June 10, 1933.
2. Erwin Schrödinger, Nature and the Greeks, Cambridge University Press, Cambridge, England, 1954, p. 95.
3. Sir Charles Sherrington, Man On His Nature, Cambridge University Press, Cambridge, England, 1949, pp. 260-261.

CHAPTER 10: THE IDEAL AND THE ACTUAL

1. Werner Heisenberg, Physics and Philosophy, HarperCollins Publishers, New York, 1958, pp. 170-171
2. Pierre Simon Laplace, A Philosophical Essay on Probabilities, John Wiley & Son, Chapman & Hall, London, England, 1902, p. 4.
3. Niels Bohr, "Physical Science and the Problem of Life," in Atomic Physics and Human Knowledge, Dover Publications, New York, 1961, p. 97.
4. Erwin Schrödinger, Nature and the Greeks, Cambridge University Press, Cambridge, England, 1954, p. 79-80.
5. Karl Menger, "The Theory of Relativity and Geometry," in Albert Einstein: Philosopher-Scientist, vol. 1, Paul Schilpp (ed.), Harper Torchbooks, New York, Harper & Row, 1949, p. 459.

CHAPTER 11: THE ATOM AND THE CELL

1. Sir Charles Sherrington, Man On His Nature, Cambridge University Press, Cambridge, England, 1949, p. 63.
2. David Bohm, Wholeness and the Implicate Order, Routledge, New York, 1980, p. 246.
3. William MacNeile Dixon, The Human Situation, Reitell Press, 2008, p. 146.
4. Erwin Schrödinger, What Is Life? With Mind and Matter, Cambridge University Press, Cambridge, England, 1944, pp. 131-132.
5. William MacNeile Dixon, The Human Situation, Reitell Press, 2008, pp. 142-143.
6. Niels Bohr, "Unity of Knowledge," in Atomic Physics and Human Knowledge, Dover Publications, New York, 1961, p. 78.
7. Erwin Schrödinger, Science and Humanism, Cambridge University Press, Cambridge, England, 1961, pp. 162-163.
8. Niels Bohr, "Atoms and Human Knowledge," in Atomic Physics and Human Knowledge, Dover Publications, New York, 1961, p. 92.
9. William MacNeile Dixon, The Human Situation, Reitell Press, 2008, p. 222.
10. Ernst Mach, The Analysis of Sensations, and the Relation of the Physical to the Psychical, 1914, pp. 98-99.
11. Niels Bohr, "Atoms and Human Knowledge," in Atomic Physics and

Human Knowledge, Dover Publications, New York, 1961, p. 92.

CHAPTER 12: DNA AND IDENTITY

1. Ernst Mach, The Analysis of Sensations, and the Relation of the Physical to the Psychical, 1914, pp. 235-236.

CHAPTER 13: THE WHOLE AND THE PARTS

1. David Bohm, Wholeness and the Implicate Order, Routledge, New York, 1980, p. 159.
2. Bernard d'Espagnat, On Physics and Philosophy, Princeton University Press, New Jersey, 2006, p. 59.
3. Osho. The Beloved, Volume 2, Rebel Publishing House, 1976, p. 116.
4. Sir Charles Sherrington, Man On His Nature, Cambridge University Press, Cambridge, England, 1949, p. 77.
5. David Bohm, Wholeness and the Implicate Order, Routledge, New York, 1980, p. 219.
6. David Bohm, Wholeness and the Implicate Order, Routledge, New York, 1980, pp. 222-223.
7. David Bohm, Wholeness and the Implicate Order, Routledge, New York, 1980, pp. 11-12.

CHAPTER 14: THE CONTINUOUS AND THE DISCRETE

1. John Bell. The Continuous and the Infinitesimal in Mathematics and Philosophy, Polimetrica, 2006, pp. 13-14.
2. John Bell. The Continuous and the Infinitesimal in Mathematics and Philosophy, Polimetrica, 2006, p. 14.
3. Henry Margenau, "Einstein's Conception of Reality," in Albert Einstein: Philosopher-Scientist, vol. 1, Paul Schilpp (ed.), Harper Torchbooks, New York, Harper & Row, 1949, pp. 257-258.
4. David Bohm, Wholeness and the Implicate Order, Routledge, New York, 1980, pp. 157-158.

CHAPTER 16: THE MEASURABLE AND THE IMMEASURABLE

1. David Bohm, Wholeness and the Implicate Order, Routledge, New York, 1980, pp. 25-26.

2. David Bohm, Wholeness and the Implicate Order, Routledge, New York, 1980, pp. 150-151.
3. David Bohm, Wholeness and the Implicate Order, Routledge, New York, 1980, p. 29.

BIBLIOGRAPHY

- Bell, John, The Continuous and the Infinitesimal in Mathematics and Philosophy, 2006.
- Bell, John, "On the Einstein Podolsky Rosen Paradox," Physics Physique Физика, 1, 1964, pp. 195–200.
- Bohm, David. Wholeness and the Implicate Order. 1980.
- Bohr, Niels. Atomic Physics and Human Knowledge, 1961.
- Capra, Fritjof. The Tao of Physics. 1975.
- d'Espagnat, Bernard. On Physics and Philosophy. 2006.
- Dixon, William Macneile. The Human Situation. 1937.
- Einstein, Albert. "Address For United Jewish Appeal." Radio broadcast, April 11, 1943.
- Einstein, Albert, Podolsky, Boris, Rosen, Nathan, "Can Quantum-Mechanical Description of Physical Reality Be Considered Complete?" Physical Review, 47, 1935, p. 777.
- Einstein, Albert. "Ernst Mach." Physikalische Zeitschrift 17 (1916): 101, 102.
- Einstein, Albert. "On the Method of Theoretical Physics." The Herbert Spencer Lecture, delivered at Oxford, June 10, 1933.
- Einstein, Albert. Physics and Reality, 1936, pp. 351-352.
- Heisenberg, Werner. Physics and Philosophy, 1958.
- Laplace, Pierre Simon. A Philosophical Essay on Probabilities. 1902.
- Mach, Ernst. The Analysis of Sensations, and the Relation of the Physical to the Psychical. 1914.
- Maxwell, J. Clerk. "A Dynamical Theory of the Electromagnetic Field," Philosophical Transactions of the Royal Society of London,

Vol. 155 (1865), pp. 459-512.

- Osho. The Beloved, Volume 2. 2001.
- Planck, Max. The New Science. 1959.
- Schilpp, Paul Arthur. Albert Einstein: Philosopher-Scientist. 1949.
- Schrödinger, Erwin. Nature and the Greeks. 1954.
- Schrödinger, Erwin. What Is Life? With Mind and Matter. 1944.
- Schrödinger, Erwin. Science and Humanism, 1961.
- Sherrington, Sir Charles. Man On His Nature. 1949.

www.ingramcontent.com/pod-product-compliance
Lightning Source LLC
Chambersburg PA
CBHW060914120626
46553CB00001B/320